The Language of nature in Buffon's *Histoire naturelle*

OXFORD UNIVERSITY STUDIES IN THE ENLIGHTENMENT – formerly *Studies on Voltaire and the Eighteenth Century* (SVEC), is dedicated to eighteenth-century research.

http://www.voltaire.ox.ac.uk/our-books/for-authors-studiesenlightenment

The Voltaire Foundation is a department of the University of Oxford. It furthers the University's objective of excellence in research, scholarship and education by publishing worldwide.

The Language of nature in Buffon's *Histoire naturelle*

HANNA ROMAN

Published by Liverpool University Press on behalf of

ISBN 978 1 78694 139 8
Oxford University Studies in the Enlightenment 2018:10
ISSN 0435-2866

Voltaire Foundation
99 Banbury Road
Oxford OX2 6JX, UK
www.voltaire.ox.ac.uk

A catalogue record for this book is available from the British Library

The correct style for citing this book is
Hanna Roman, *The Language of nature in Buffon's* Histoire naturelle,
Oxford University Studies in the Enlightenment
(Liverpool, Liverpool University Press, 2018)

Cover illustration: *histoire naturelle. Premier discours de la manière d'étudier et de traiter l'Histoire Naturelle*, from *Illustrations de Histoire naturelle générale et particulière avec la description du Cabinet du Roy.*
Source: gallica.bnf.fr / Bibliothèque nationale de France

Printed and bound in Poland by BooksFactory.co.uk

Oxford University Studies in the Enlightenment

The Language of nature in Buffon's *Histoire naturelle*

Drawing from literary studies, philosophy and the history of science, in this interdisciplinary study Hanna Roman argues that the language of Buffon's *Histoire naturelle* (1749-1788) could not be separated from the science it conveyed; the language communicated nature's vital order, form and movement. In the *Histoire naturelle*, the ability of language to embody and communicate the living essence of nature grew increasingly poignant as Buffon established his hypothesis that the Earth, initially a molten ball of fire, was dying as it slowly became colder.

The author highlights Buffon's 'Epoques de la nature' (1778) in which he implied that to save nature from cold death, people must learn to create actual heat according to the model provided by his lyrical, dynamic language, the energy of which would transform into re-warming a cooling globe.

In this way, Roman argues that Buffon's literary simulacrum of nature taught his readers not only about the history of nature and its laws, but also how to interact with nature differently, transferring to them the skills necessary to modify the surrounding world in order to better fit the desires and dreams of humanity. A new world could be more than imagined – it could be engineered through language.

This book is dedicated to Dr Wilda Anderson
and to Dr David Harrison,
for their inspiration and guidance.

Contents

Preface

> Quelle différence y auroit-il entre la lecture d'un ouvrage où tous les ressorts de l'univers seroient développés, & l'étude même de l'univers? presqu'aucune.
>
> What difference would there be between the reading of a work in which all the motives of the universe are explored, and the very study of the universe? Almost none.[1]

> Remember that the authors need the reader's willingness to have their own claims turned into facts [...]. If the readers are put off, they are not going to take up the claim; but if they are left free to discuss the claim, it will be deeply altered. The writer of a scientific text is then in a quandary: how to leave someone completely free and have them at the same time completely obedient. What is the best way to solve this paradox? To lay out the text so that wherever the reader is there is only *one way* to go.[2]

This project found its beginnings in the confusion that confronts modern readers as they try to make sense of the *Histoire naturelle*

1. Denis Diderot, 'Encyclopédie', in *Encyclopédie, ou Dictionnaire raisonné des sciences, des arts et des métiers*, ed. Denis Diderot and Jean D'Alembert, ARTFL *Encyclopédie* project, ed. Robert Morrissey, http://encyclopedie.uchicago.edu/ (last accessed March 14, 2018), vol.5, p.641. Unless otherwise noted, all translations from the French in this book are my own.
2. Bruno Latour, *Science in action: how to follow scientists and engineers through society* (Cambridge, 1988), p.57 (original emphasis).

of the Enlightenment natural historian Georges-Louis Leclerc de Buffon. There are two disparate views of Buffon. On the one hand, he attempted to do science and seems to have gotten it wrong. His theories about Earth history and biology are known to have contributed to the sciences of geology and evolution in the nineteenth century. Yet many of them were discredited – even by Buffon's peers in his own century. On the other hand, Buffon wrote beautifully. When remembered in the centuries after his death, it was usually for the imposing and sonorous quality of his language. As Georges Cuvier wrote in his biographical notice, there were many opinions on Buffon as a natural historian, but

> [i]l n'y a qu'une opinion sur Buffon considéré comme écrivain: pour l'élévation du point de vue où il se place, pour la marche forte et savante de ses idées, pour la pompe et la majesté de ses images, pour la noble gravité de ses expressions, pour l'harmonie soutenue de son style dans les grands sujets, il n'a peut-être été égalé par personne.

> [t]here is only one opinion about Buffon, considered as a writer: for the elevated point of view from which he speaks, for the strong and learned progression of his ideas, for the ceremony and majesty of his images, for the noble gravity of his expressions, for the formal harmony of his style [when writing about] grand subjects, he was perhaps not equaled by anyone.[3]

It is difficult not to focus entirely on Buffon's language and imagine that this, and not the ideas behind it, was the primary focus of his work. Wrestling with the apparent disparity between the fact that Buffon was a well-known scientific figure in the Enlightenment with influential ideas and the impression that Buffon's poetic language is the only aspect of the *Histoire naturelle* that has survived the test of time, the modern naturalist Stephen Jay Gould writes that 'Buffon had a passion for order, but he developed no central theory that could be defined by a memorable phrase or concept. He wrote volumes of incomparable prose and propagated ideas, sometimes quite radical, about all major subjects in natural history. But no central thread unites his system.'[4] In the chapters that follow, I demonstrate quite the opposite: that the 'central thread' of the *Histoire naturelle* was indeed its

3. Georges Cuvier, 'Buffon', in *Biographie universelle ancienne et moderne*, 85 vols (Paris, 1812), vol.6, p.236-37.
4. Stephen Jay Gould, 'Inventing natural history in style', in *The Lying stones of Marrakech: penultimate reflections in natural history* (New York, 2000), p.75-90 (83).

written language, its 'incomparable prose', which formed the basis for knowledge of nature, past, present and future. As writers of nature, Gould and Buffon are similar. Both convey the technical structure of the material world through figurative language, recreating nature in readers' minds and thereby allowing these readers to better know, speak and act in nature's future.

This book argues that written language was not simply an embellishment or a vulgarization of a system of facts to make it more appealing or convincing to a larger public. Writing was the foundation of this system itself. As the basis or underlying mechanism of knowledge of the natural world, language was the tool that Buffon used to redevelop the discipline of natural history. Through techniques of written language, this new discipline was defined as the harmonization of natural phenomena and events with the human perception of them. That the science of nature was explained through its representation in discourse was an advantage, and not a hindrance, to Buffon, who held that the facts of the world became more true as people learned to better read and express them. Nature and language were deeply related for Buffon; to study him as either a literary or a scientific figure would be to ignore the fundamental connectivity he placed between these two realms of thought.

Why is Buffon still important today, and in what domains did his work have a lasting impact? The significance of the *Histoire naturelle* continues to echo in the idea that the language in which a form of knowledge is conveyed is as important as the knowledge itself. This idea was at the foundation of the liberal arts curriculum and the European university, dating back to the Middle Ages: one had first to learn the arts of speaking – grammar, rhetoric, dialectic – before it could be possible to study and articulate the structure of the natural world and the cosmos. It continues to be important in the modern university: without comprehending the matrix of words and ideas that composes a language, historical processes, cultural identities and scientific advances become unintelligible. Science continues to be not a static collection of facts but an active, growing body of knowledge that sets up models through which to interrogate reality and eventually to get closer to it. Language – mathematical, theoretical, descriptive, hypothetical – forms that bridge between the human mind and the world. If this book focuses specifically on one figure of Enlightenment science, its larger meaning demonstrates the importance of conducting a humanistic, literary investigation of a moment in the past, and reveals the continued importance of studying science and literature side by side.

Introduction: Enlightenment natural history and literary invention

> INVENTER [...]. Trouver quelque chose de nouveau par la force de son esprit, de son imagination. *Inventer un Art, une Science. Inventer un systême, une machine.*
>
> TO INVENT [...]. To find something new through the force of one's mind, of one's imagination. *To invent an Art, a Science. To invent a system, a machine.*[1]

Georges-Louis Leclerc, comte de Buffon (1707-1788), is known as one of the most famous, prolific and articulate naturalists of Enlightenment France. A member of both the Parisian Académie royale des sciences and the Académie française, he was also the intendant of the Jardin du roi, from 1739 until his death in 1788, collecting specimens of plants and animals and other treasures from around the world. He is best known for his thirty-six volumes of the *Histoire naturelle*, published between 1749 and 1788, which addressed a wide variety of themes, from taxonomy and nomenclature, to geology and the origins of the Earth, to embryology and questions of race and species identity. Among Buffon's other recognized works are the much admired and frequently cited 'Discours sur le style' (1753) and one of his last and most passionate works, the 'Epoques de la nature' (1778).

What does it mean, to call Buffon a natural historian, and to identify him with the discipline of natural history? An important aspect of this identification is that he was not only part of the Académie des sciences

1. 'Invention', in *Dictionnaire de l'Académie française*, 4th edn (Paris, 1762), ARTFL *Dictionnaires d'autrefois*, ed. Robert Morrissey, http://artfl-project.uchicago.edu/content/dictionnaires-dautrefois (last accessed March 15, 2018).

but also a member of the Académie française, for which the 'Discours sur le style' was his acceptance speech. The latter is an institution now associated with the study of literature and written language, not with the 'connected body of demonstrated truths or [...] observed facts systematically classified and more or less comprehended by general laws, and incorporating trustworthy methods (now esp. those involving the scientific method and which incorporate falsifiable hypotheses) for the discovery of new truth in its own domain', as the term 'science' is defined in the modern *Oxford English dictionary*.[2] Linking facts and truths to the larger laws of nature and discovering the methods by which to make this linkage were, however, the work of ordered ideas and written language, or what Buffon called 'style'. In this book, I will use the term 'style' in this broader sense in which Buffon defined it.

Style: combining rhetoric and knowledge

Today, style has at least two meanings: it can be the manner in which an individual author writes, or it can refer to a more general movement in literature, art or fashion. If style is a literary idea referring to the creativity and individuality of an author, why did Buffon place so much emphasis on the natural world when discussing style? In the 'Discours sur le style' he asked, 'Pourquoi les ouvrages de la Nature sont-ils si parfaits?' and affirmed that if a writer 'imite la Nature dans sa marche et dans son travail [...] il établira sur des fondemens inébranlables, des monumens immortels'.[3] He underlined the point that '[l]e ton du Philosophe pourra devenir sublime toutes les fois qu'il parlera des loix de la Nature, des êtres en général, de l'espace, de la matière, du mouvement et du temps, de l'ame, de l'esprit humain, des sentimens, des passions.'[4] Buffon seems to have understood himself to

2. 'Science, n.', in *Oxford English dictionary online*, http://www.oed.com.eres.library.manoa.hawaii.edu/view/Entry/172672?redirectedFrom=science (last accessed March 15, 2018).
3. 'Why are the works of Nature so perfect?'; 'imitates the operation and work of Nature [...] he will establish, upon unshakable foundations, immortal monuments.' Georges-Louis Leclerc, comte de Buffon, 'Discours sur le style', in *Histoire naturelle, générale et particulière: supplément*, vol.4 (Paris, Imprimerie royale, 1777), p.1-13 (5-6).
4. 'The Philosopher's tone can become sublime whenever he speaks of the laws of Nature, of beings in general, of space, of matter, of movement and of time, of the soul, of the human spirit, of feelings, of passions.' Buffon, 'Discours sur le style', p.12.

be both a natural historian, observing the world, and an orator, seeking to communicate these observations in the most persuasive (but not necessarily or absolutely true) manner possible. In the 'Discours sur le style' he stated that the disciplines of political history and philosophy could become sublime when speaking of great thinkers or of natural laws, but in general their tone should not endeavor to be too grand. This rule changed when discoursing on the relationship between people and nature. This language, attributed to orators and poets (and, Buffon hoped, to the natural historian), had to be sublime.

The task of the orator was essential to, and a distinctive aspect of, Buffon's definition of the subject area of natural history. Natural history was not just about understanding how the visible world was structured. It also entailed delving into what Buffon called the 'archives' of a distant geological past and piecing together events that were unobservable to the human eye. The story of nature could not merely be recounted, but had to be told both convincingly and persuasively, in a way that would appeal both to a reader's reason and prior knowledge and to his or her recognition of nature's and language's sublime beauty and power. This implied thinking not only about the rhetoric and the register of language, but also about a different way of perceiving the reading public – which was only beginning to be assumed and tested by natural philosophers (and other authors) of the time. Many works of natural history, astronomy, physics and chemistry of the period were still circulated in private networks of well-informed correspondents, and were not necessarily destined for wider dissemination, let alone for publication. Buffon's works were designed to speak to a larger public and sought its active participation. They were published, read within many circles, admired, debated and heavily criticized. His texts reflect a desire to convince and unite an ideal community of readers with varying perspectives and understandings through the same method and argumentative structures.

The *Histoire naturelle* was built upon an underlying theory of how the human mind absorbed and assimilated knowledge, which determined how this knowledge should be presented, organized and disseminated. From Buffon's point of view, natural history was the ideal space in which the process of transmission from world to idea to written word could be investigated and tested, and its implications for the future of the relationship between human thought and nature imagined. The 'history' entails both events that happened in the past and the written narration and interpretation of these events. For Buffon, the history of nature embodied events and discourse:

events were determined and only became knowable through human understanding of and belief in them.

Buffon's *Histoire naturelle* is thus an example of how, in the Enlightenment, language was a way of knowing, inseparable from the information it imparted. The epistemological strength of the *Histoire naturelle* derived from the ability of its language to convincingly and powerfully reproduce the structure and activity of nature, to become nature itself in a written form. In order to comprehend the *Histoire naturelle* and its implications, scientific and literary, this work needs to be reinserted into the intellectual and discursive assumptions of its time, assumptions that took the written word to be part of the process of generating knowledge. Language was the embodiment of method, a reflection of and monument to the construction of a science. As the abbé de Condillac wrote, 'Toute langue est une méthode analytique et toute méthode analytique est une langue.'[5] For Buffon, language was itself *scientia*, in that it gave order and demonstrability to knowledge of the world and endowed its author, and subsequently its ideal readers, with the capacity to investigate, understand and ultimately intervene in the work of nature. The more people learned to read and interpret nature, the more involved they were in its story. The history of nature and the history of human thought became deeply intertwined in the *Histoire naturelle* through a shared narrative of reshaping the face of the planet Earth. In this way, interpreting Buffon's work through the activity of his language enables its contemporary value as a serious scientific and humanistic object to be more accurately gauged. This interpretation allows for a more holistic conception of the rational and poetic goals of Enlightenment natural history.

Style, in its modern sense of an author's distinctive characteristics, was but one element of Buffon's larger intellectual and discursive project. It was one important part of a larger understanding of written language as itself constitutive of natural knowledge. This book examines and relates the different aspects of written language in Buffon's *Histoire naturelle*, tracing how they developed consubstantially alongside his ideas about nature. His work makes evident the argument that natural philosophical writing in the Enlightenment constructed the human experience of the world, such that the natural became as much a fact of the outside, material realm as an invention of

5. 'Every language is an analytical method and every analytical method is a language.' Etienne Bonnot de Condillac, *La Langue des calculs*, ed. Sylvain Auroux and Anne-Marie Chouillet (Lille, 1981), p.1.

human perception. By breaking down the conception and composition of the *Histoire naturelle* into its successive aspects or stages, from apprehending nature's order in the mind to actively changing this order in the surrounding environment, this book demonstrates that the natural world was known through the language that expressed it, and that there could be no science of natural history without written language. Each stage of the development of Buffon's discipline of natural history is marked by a chapter that analyses a discursive and epistemological aspect of Buffon's thought – *mathesis* (the method of generating rational thought), natural law, *energeia*, history, hypothesis, heat. Scaling these levels of comprehension of the world, it becomes clear that the cohesive, operative and useful body of knowledge called natural history could only be produced through language.

The chapters that follow dissect the philosophical, epistemological and discursive processes through which human understanding and language coalesced with the history and operation of nature in the *Histoire naturelle.* Through a study of the relationship between universal mathematics and natural law, the first chapter examines how nature was observed, ordered and systematized in the beginning of the eighteenth century. The second chapter turns to the idea of heat, a key rhetorical *and* epistemological tool for Buffon, which he used to render the system of nature visible and palpable to the mind through a language that had both poetic and material force. Only after an analysis of perception and heat can the composition of natural history be understood, as the harmonization of language and world, and as the creation of a dual sense of nature in which the physical, or the essence of the material world, was merged with the moral and historical, or the record of human thought and activity (Chapter 3). The fourth chapter introduces Buffon's use of hypothesis: the idea that people could use language not only to capture natural order and operation, but also to project into the domain of the probable, to speculate on events and activities not yet observed. It was through this ability to reproduce nature's power in language that Buffon believed human beings could transform their imagination of the future of the world into reality, a vision discussed in the final chapter. Human control of natural environments was, at its very core, defined by the deeply literary endeavor of inventing and controlling the future of nature through language.

Harmonizing world and word

How were science and language related in Buffon's work? To answer this question, I start, briefly, at the end of Buffon's oeuvre, introducing the curious idea of learning to control, manage and ameliorate natural environments through language. In the 'Epoques de la nature' (1778), he told the story of the geological development of the planet Earth, over the course of seven chapters, or 'epochs'. He famously postulated that Earth and the other known planets were once pieces of the Sun, torn off of this star by the force of a comet's collision. The world began as a molten ball of fire. Over the course of historical time and over the course of his book, Buffon traced its slow cooling. (His narrative of the Earth's geological and natural history extended the age of the planet far deeper into the past than was commonly accepted at the time.) His argument implied that the depletion of the planet's heat would eventually lead to its inevitable end as a frozen globe of ice – not a religious apocalypse, but a natural one, in historical time. In the last chapter of the 'Epoques', Buffon reiterated this dire prediction, but then added an unexpected twist, attempting to convince his readers of their ability to slow down the Earth's cooling by producing heat: 'Rien ne paroît plus difficile, pour ne pas dire impossible, que de s'opposer au refroidissement successif de la Terre et de réchauffer la température d'un climat; cependant l'homme le peut faire et l'a fait [...]. Assainir, défricher et peupler un pays, c'est lui rendre de la chaleur pour plusieurs milliers d'années.'[6]

By the end of the sixth and penultimate epoch, whatever heat was left in the Earth had to be maintained, and yet nature was becoming less and less capable of doing so on its own. It would soon no longer enough for people to know nature, passively receiving the word of the author of natural history. Buffon's language now also had to communicate the ability to act upon nature. The rhetorical impact of his text had to be translated into a physical force that could change peoples' relationship to their environment and teach them to modify their surroundings. Their manipulation of nature became essential

6. 'Nothing would seem more difficult, if not to say impossible, than to counter the successive cooling of the Earth and to reheat the temperature of a climate: yet man can and has done it [...]. To clear, cultivate and populate a country, is to give it heat for several thousand years.' Buffon, 'Des époques de la nature', in *Histoire naturelle, générale et particulière: supplément*, vol.5 (Paris, Imprimerie royale, 1778), p.1-254 (238-40).

in the 'Septième époque: Lorsque la puissance de l'homme a secondé celle de la Nature' ('When the power of man assisted that of Nature'), in order to produce the heat that nature could no longer make. This final chapter of the 'Epoques' seemed to imply that, by reading Buffon, people could learn to imitate nature's power over the planet, ultimately creating what Buffon earlier had termed 'a new Nature' ('une Nature nouvelle'), through physical modification of the world.[7] This chapter reads to modern eyes like a defense of global warming: it seems that only human industry – in those days, clearing forests and swamps and cultivating the land – could rewarm and rescue a dying world. However, this would not have been considered in a negative light, as is man-made climate change today, but rather as another example of the Enlightenment's hopes for the realization of the power of the human spirit to triumph over natural law and create its own laws in a future, conceivably better, world.[8]

Over the course of the *Histoire naturelle*, Buffon used language to measure the development both of knowledge and of human ability with respect to the creative power of nature. In the 1740s and 1750s, early in his career, Buffon did not write of the cooling of the Earth, but he did speak of heat in a rhetorical and metaphorical sense, describing the energetic movement through which ideas were transferred between author and reader. As it developed, his natural historical project was increasingly grounded in the idea that knowledge became scientific and authoritative when the discourse of thought, that is, the narrative of the order and structure of his ideas, mirrored the order and structure of the world itself. This simulacrum of nature was translated for readers through a language imbued with what he called 'chaleur' ('heat'), or a persuasive, poetic discourse that captured the order and vitality of nature. To produce knowledge accepted as certain and authoritative – scientific knowledge – Buffon believed that language about nature had to be structured in the same manner, and have the same dynamic liveliness, as the outside world. He spoke of the energetic transfer of ideas as part of the process of painting with words. Indeed, when writing, and especially when writing about the natural world, he insisted in the 'Discours sur le style' that 'il ne suffit

7. Buffon, *Histoire naturelle, générale et particulière, avec la description du Cabinet du roy*, 15 vols (Paris, Imprimerie royale, 1749-1767), vol.12 (1764), p.xii-xiii.
8. On the Enlightenment belief in the power of the human mind, see Jean Dagen, *L'Histoire de l'esprit humain dans la pensée française de Fontenelle à Condorcet* (Paris, 1977).

pas de frapper l'oreille et d'occuper les yeux; il faut agir sur l'âme et toucher le cœur en parlant à l'esprit.'[9] Buffon's language of nature became the translative node between raw observation and knowledge of nature, shaping this knowledge out of both the sensory and the rational components of experience.

In the *Histoire naturelle*, Buffon's 'new nature' is recognizable because it resembles the original nature of the outside world. To borrow a term from Nelson Goodman and the history of art, the nature portrayed in Buffon's language is 're-cognizable' from the order of the outside world.[10] The 're-cognition' of nature implies the epistemological and poetic processes of moving between and harmonizing the related worlds of reality and perception. Furthermore, it is another way of explaning Buffon's intention that his new, imagined world ask readers to reconsider and reimagine their role in the world currently surrounding them, considering how reality might be altered and improved according to the story they choose to tell about it.[11] This idea of re-cognition, the basis of 'worldmaking', addresses the discursive value of the *Histoire naturelle* as an instrument for modifying perception and judgment of reality, an aspect of the development of the Enlightenment philosophical, critical spirit which I see as important to understanding the role of literature at this time. Read this way, the *Histoire naturelle* becomes both scientific and literary, an experiment in examining nature as it was and as it is in order to imagine how it could be modified. It reflects a creative optimism about how, perhaps, nature would someday become the realization of human imagination.[12]

Goodman furthermore argues that 'the arts must be taken no less seriously than the sciences as modes of discovery, creation, and enlargement of knowledge in the broad sense of advancement of the understanding.'[13] He discusses how reality is the product of the

9. '[I]t is not enough to strike the ear and to occupy the eyes; it is necessary to act upon the soul and to touch the heart by speaking to the mind.' Buffon, 'Discours sur le style', p.3.
10. Nelson Goodman, *Ways of worldmaking* (Indianapolis, IN, 1978), p.105.
11. On the literary, fictional nature of Buffon's 'Epoques de la nature', see Benoît de Baere, 'Une histoire imaginée … mais vraie: le problème du statut des *Epoques de la nature* de Buffon', in *Les Songes de Clio: fiction et histoire sous l'Ancien Régime*, ed. Sabrina Vervacke, Eric Van der Schueren and Thierry Belleguic (Québec City, 2006), p.247-62 (254).
12. A question raised in Gillian Beer, 'Has nature a future?', in *The Third culture: literature and science*, ed. Elinor S. Shaffer (Berlin, 1998), p.15-27 (16).
13. Goodman, *Ways of worldmaking*, p.102.

ways in which literary, artistic and scientific 'worlds' have trained people to perceive. 'Worldmaking' – be it in the realm of fiction, nature, aesthetics, history, politics – is the poetic invention of new and productive ways of reading and understanding reality. This argument is a helpful lens through which to understand the conception and production of scientific knowledge in eighteenth-century France. It is even more applicable if Goodman's words are turned around: in order to understand bodies of knowledge in the Enlightenment as scientific, they should be studied in the same manner as literature, in which new, invented realms are proposed and built through 'modes of discovery', or the framework in which an author makes the rules for a world and orders ideas and events therein.

Natural history: between physics and history

'Scientific knowledge is not just any knowledge. It is useful and *necessary* knowledge', Wilda Anderson has written. 'There is a natural harmony between the processes of science and the physical structure of the world. The one was designed in order to allow us to perceive the other.'[14] In the Enlightenment, science could be described as tantamount to the narrative of the discovery of nature's structure, the language of nature's laws and processes. In the *Dictionnaire de l'Académie française* 'science' was defined as the general 'connoissance qu'on a de quelque chose' ('knowledge that one has of something'). One way of knowing scientifically was the 'connoissance certaine & évidente des choses par leurs causes' ('certain and evident knowledge of things through their causes').[15] The certainty and evidence of knowledge was determined by the method by which it was produced, and how this method proposed to link the perceived order of the world in the mind with the real order of the world itself (a question further explored in Chapter 1). Method gave way to language, and written language became the monument or physical trace of this process of making knowledge. Language formed the tangible bridge between mind and world.[16] To study a

14. Wilda C. Anderson, *Between the library and the laboratory. the language of chemistry in eighteenth-century France* (Baltimore, MD, 1984), p.12 (original emphasis).
15. 'Science', in *Dictionnaire de l'Académie française*, n.p.
16. As Michel Foucault wrote, natural history worked 'to bring language as close as possible to the observing gaze, and the things observed as close as possible to words' (*The Order of things: an archaeology of the human sciences*, New York, 1970, p.132).

body of knowledge was thus to study the language that created it: in L. J. Jordanova's words, '[w]ays of writing about nature [...] [were] bound up with ways of knowing about nature.'[17] Moreover, in the eighteenth century language was not thought to be solely produced by the human mind; it was understood to be an inherent part of the world as well, and it required a natural philosopher's talent and sensibility to read and interpret the book of nature, and to extract and translate it into a format accessible to the human intellect.[18] It was not simply that new knowledge of nature changed the ways in which nature was represented and discussed. Discourses of nature likewise were designed to alter how the world was perceived and how people interacted with it.

Buffon, in particular, did not define the science of natural history as simply the study of order of the world, but furthermore sought to emulate in written words – and eventually appropriate, through human activity – what he believed to be the principles and mechanics of natural creation. It was the alternate, textual nature, sculpted through the vitality of Buffon's language, that likely caused subsequent centuries to read his work as unsubstantiated by fact, lacking rigor and analysis, and making light of or vulgarizing the serious scientific ideas of his day for the enjoyment of popular audiences and for his own fame. This reading of Buffon divorces the *Histoire naturelle* from what was seen as the more scholarly work of his contemporaries in mechanics, physics and natural history such as D'Alembert, Clairaut and Linnaeus. Buffon was depicted as a rhetorician and a poetic precursor to the study of the one, denotative reality that purportedly came to define modern science.[19] Until well into the twentieth century, Thierry Hoquet has shown, Buffon was considered in a negative light with respect to what was considered real, productive scientific work. Hoquet writes that Buffon was depicted as leading the cause for scientific vulgarization, 'une science mondaine pétrie de la curiosité' ('a science of society, inflated with curiosity'); he was described

17. L. J. Jordanova, 'Introduction', in *Languages of nature: critical essays on science and literature*, ed. L. J. Jordanova (London, 1986), p.15-47 (28).
18. On the concept of reading nature as a book, see Ernst Robert Curtius, *European literature and the Latin Middle Ages*, translated by Willard R. Trask (Princeton, NJ, 1973), p.320-24.
19. Gaston Bachelard gives Buffon as an example of a 'pre-scientific mind' ('esprit préscientifique'), whose work develops from subjective experience of the world and hence precedes the advent of what is now called science. See Gaston Bachelard, *La Formation de l'esprit scientifique: contribution à une psychanalyse de la connaissance objective* (Paris, 1960), p.23-25.

as a 'grand poète de la Nature' ('great poet of Nature'), 'génial et solitaire' ('brilliant and solitary'). Some of his famous works such as the 'Discours sur le style' and the 'Epoques de la nature' were later published separately. They were read and commented upon as literary works. Removed from the larger context of the *Histoire naturelle*, they came to be considered beautiful, poetic descriptions of nature by 'un naturaliste encore inabouti' ('a yet unaccomplished naturalist') – as Hoquet captures the spirit of Buffon's later critics.[20] Furthermore, upon Buffon's death in 1788, the structure, aims and tone of natural science were beginning to change. The study of nature would gradually be split into more specialized disciplines and fields, especially as the French Revolution demanded projects and institutions that would work for the good of the country by advancing military technology or food production, for example.[21] Historians have noted a shift in focus away from Buffon's quest for the most general laws of nature accompanied by a language many were starting to see as literary and fanciful, to a 'need for geometrical precision and rigor' in examining the world in all its details.[22]

To read the *Histoire naturelle* either as an exercise in poetic language or as a more or less successful attempt to produce accurate knowledge about nature is, however, not only to misunderstand how this work operates, but to reduce a complex process of knowledge-making to disciplinary boundaries that did not exist in the Enlightenment. Buffon worked and rose to fame and prominence – both intellectual and popular – at a time when the definition of natural history was in flux, consisting of competing epistemologies and practices and appealing in varied ways to philosophers and academics, collectors and fashionable society.[23] Because of this wide range of interests and

20. Thierry Hoquet, *Buffon: histoire naturelle et philosophie* (Paris, 2005), p.12-15.
21. Pietro Corsi, *The Age of Lamarck: evolutionary theories in France, 1790-1830*, translated by Jonathan Mandelbaum (Berkeley, CA, 1988), p.10.
22. Corsi, *The Age of Lamarck*, p.24. On the reception of Buffon's work after his death, see the concluding chapter of the present volume.
23. On natural history in the Enlightenment, see Emma C. Spary, 'The "nature" of Enlightenment', in *The Sciences in enlightened Europe*, ed. William Clark, Jan Golinski and Simon Schaffer (Chicago, IL, 1999), p.272-304; *Cultures of natural history*, ed. N. Jardine, J. A. Secord and E. C. Spary (Cambridge, 1996); Daniel Roche, 'Natural history in the academies', in *Cultures of natural history*, ed. N. Jardine, J. A. Secord and E. C. Spary, p.127-44; and Roger Hahn, *The Anatomy of a scientific institution: the Paris Academy of Sciences, 1666-1803* (Berkeley, CA, 1971). On the sources which influenced Buffon, see Hoquet, *Buffon*, ch.1 and 3.

publics, authors of natural history paid close attention to the rhetoric and language of their texts, to how arguments were presented and to whom – especially as the number of readers of natural history grew over the course of the century.[24] The foundational studies by Jacques Roger, Thierry Hoquet and Jeff Loveland on Buffon's work and philosophy have shown how Buffon reconsidered, combined or rejected existing discursive modes of natural philosophy to carve out a new space which reconsidered nature's relationship to history and time and thus to perception and language.[25] They show how both previous and contemporary ideas and debates within France's intellectual spheres shaped and influenced Buffon's philosophy of nature and what he believed should be presented in the text. They importantly place the style – in this case, his manner of expression – of Buffon's texts into the context of the practices typical of both rhetoric and natural history writing at the time.[26]

However, these works do not go far enough in demonstrating the fundamental connection between written language and knowledge in the *Histoire naturelle*. They do not fully address the question of how language was used to create meaning in the *Histoire naturelle*, that is, how it was not simply a vehicle for conveying knowledge in a lyrical and entertaining manner but indeed a form and a practice of natural knowledge itself. Joanna Stalnaker's work has been crucial in changing this interpretation of Buffon. Her insightful critique emphasizes the importance of adding complexity to or challenging the traditional separation of literary and scientific disciplines that causes expressive texts such as Buffon's to be read either as literature, which 'should not be interpreted in terms of its connections to reality,' or as science, and thus using language as '"mere rhetoric" to enliven its ideas'.[27] Her statement that 'Buffon's essay on style was influential precisely because

24. For the rhetorical figures, practices and traditions upon which Buffon drew to address different readerly expectations, see Maëlle Levacher, 'Les lieux communs dans l'*Histoire naturelle* de Buffon: rhétorique judiciaire, rivalité dans la récriture et compromis épistémologique', *Dix-huitième siècle* 42 (2010), p.35-56; M. Levacher, *Buffon et ses lecteurs: les complicités de l'"Histoire naturelle'* (Paris, 2011); and Jeff Loveland, *Rhetoric and natural history: Buffon in polemical and literary context*, *SVEC* 2001:03, p.7-8.
25. Jacques Roger, *Buffon: un philosophe au Jardin du roi* (Paris, 1989); Hoquet, *Buffon*; Loveland, *Rhetoric and natural history*.
26. See Loveland, *Rhetoric and natural history*, p.18-28.
27. Joanna Stalnaker, *The Unfinished Enlightenment: description in the age of the 'Encyclopédie'* (Ithaca, NY, 2010), p.4-5.

it made style inextricable both from the writer's individual perspective and from the world he sought to represent' is a key point of departure for my work.[28]

To understand how the *Histoire naturelle* operated as a work of knowledge about nature, it is thus crucial to consider how Buffon's project took shape within a moment of changing definitions and shifting disciplinary boundaries in the realms of physical sciences and natural philosophy, history and *belles-lettres* and the uses of language in the representation of knowledge. Traditionally, the sciences of the Enlightenment, following the Scientific Revolution, have been viewed, in the words of Richard Yeo, as the specialization of knowledge, using a modern definition of physics as a model: 'the consolidation of inquiry into the phenomena of electricity, magnetism, and heat, the revolution in chemistry, historical theories of the solar system, and the appearance of new subjects such as geology, biology, and psychology'. However, Yeo continues,

> it is precisely these developments, which appear reassuringly 'modern', that made it crucial to resist any easy importation of later disciplinary categories into the discussion of eighteenth-century science [...]. This explosion of knowledge [...] strained the old terminology and some of the classifications it embodied. It made new maps of knowledge necessary, while at the same time making them difficult to draw.[29]

It is important to remain aware of these changes in the ordering and valorization of forms of knowledge, avoiding the tendency to imagine the Enlightenment (or any historical moment) as a period when there were no disciplinary boundaries, when natural philosophers seemed to know and study everything, when the sciences were in some sort of pre-developmental stage, mixing freely with the poetic confusion of literature. Language was seen as being essential to knowledge of both physics and history. The following pages, as well as the next chapter, will show how the *Histoire naturelle* used language to strategically combine and rethink disciplinary maps and methods.

In the classification of the physical sciences, knowledge based upon the observation and interpretation of natural phenomena fell into at least two general categories in the early Enlightenment, drawn to a large extent from Francis Bacon's ordering of the sciences

28. Stalnaker, *The Unfinished Enlightenment*, p.33-34.
29. Richard Yeo, 'Classifying the sciences', in *The Cambridge history of science*, vol.4: *Eighteenth-century science*, ed. Roy Porter (Cambridge, 2003), p.241-66 (242).

in the early seventeenth century: natural philosophy and natural history. Yeo summarizes the distinction between these two: '[t]he former, located under the faculty of Reason and part of philosophy, embraced all the mathematical and physical sciences – disciplines that eighteenth-century writers recognized as the Newtonian sciences. In contrast, natural history belonged to Memory and was charged with producing adequate descriptions (histories), collections and taxonomies of minerals, plants, animals, and, significantly, accounts of manual crafts and machines.'[30] Indeed, physics encompassed a multiplicity of practices and principles that were difficult to situate in any one contemporary division of a tree of knowledge. It could refer to sciences that used mathematics such as astronomy, but it could also embrace parts of botany, anatomy, chemistry and medicine.[31] What Buffon called 'natural history' was as much 'reason' as 'memory'. It included the description and collection of natural objects, the search for causes and general laws in the natural world, and the application of mathematical reasoning. It thus had a systematic order and logic not typically associated with older practices of natural history.

As Buffon worked within older and existing categories of knowledge, combining and renegotiating them to build a new discipline for studying the natural world, he similarly had to contend with the epistemological status and uses of history. 'History' was a wide-ranging term at the time, one that, in Jean-Paul Sermain's words, 'ignore les frontières entre les modes de discours, puisqu'elle concerne des narrations vraies et d'autres fictives, ce qui s'est passé autrefois et ce qui est contemporain'.[32] For instance, a history could refer to an official historiography of the events and achievements of the reign of a king, an epic medieval romance and adventure, a novel or story based upon a historical moment (e.g. *La Princesse de Clèves*) or an account of the characteristics and qualities of an animal or plant based upon observations and the authority of previous writers dating back to Antiquity. Far from being unrelated to the natural sciences, history had long influenced them, particularly medicine, because

30. Yeo, 'Classifying the sciences', p.254.

31. Georges-Louis Leclerc de Buffon, *Histoire naturelle, générale et particulière, avec la description du Cabinet du roy*, ed. Stéphane Schmitt and Cédric Crémière (Paris, 2007-), 'Introduction', vol.7 (2011), p.7-90 (25-26).

32. '[I]gnores the boundaries between discursive modes, because it is concerned with true stories and fictitious ones, that which happened in the past and that which is contemporary'. Jean-Paul Sermain, *Le Roman jusqu'à la Révolution française* (Paris, 2011), p.27.

of its descriptive aspect. History did not have the status of natural philosophy or physics, but it was understood in the study of nature and the human body as 'the knowledge and description of particulars', according to Gianna Pomata and Nancy Siraisi. In the practice of early modern medicine, for example, case studies were called '*historiae*' or '*observationes*', and medical treatises 'regularly started with [...] a *historia*, meaning a thorough description of the structure of bodily parts, preliminary to an understanding of their function, or "use"'.[33]

Generally speaking, then, history was a description of events and things, real or imagined, witnessed directly or recounted second- or third-hand. In the eighteenth century, the *Dictionnaire de l'Académie française* defined 'histoire' as 'des actions & des choses dignes de mémoire' in the form of an 'histoire générale' or 'particulière', 'ancienne' or 'moderne', 'sacrée' or 'prophane', 'véritable' or 'fabuleuse'.[34] History was classified in the Baconian system (as well as in Ephraim Chambers' *Cyclopaedia* and in Diderot and D'Alembert's *Encyclopédie*) under 'memory', not 'reason': it was not about logical demonstration and the search for causes and laws, but rather about the judgment and interpretation of varying degrees of truth, probability and authority, and about deciding upon the rhetorical and epistemological value of a discourse within a particular context. History noted things – natural, moral and miraculous – that occurred in the world, recording these happenings without trying to conclude definitively from them. It was understood at the time, according to Donald Kelley, as 'observation, knowledge derived from personal experience or faith in the senses', as well as in human memory. Even though it contributed to the study of nature, history was often contrasted with 'science', Kelley explains, or the search for causes and the ordering of knowledge according to these causes.[35]

Despite its separation from science, and from natural philosophy in particular, history was a crucial discursive field for a thinker like Buffon because it allowed him to make the connection between the natural and the moral, the events of the world and the human

33. Gianna Pomata and Nancy Siraisi, 'Introduction', in *Historia: empiricism and erudition in early modern Europe*, ed. Gianna Pomata and Nancy Siraisi (Cambridge, MA, 2005), p.1-40 (2).
34. '[A]ctions & things worthy of memory'; a 'general' or 'particular history'; 'ancient' or 'modern' history'; 'sacred' or 'profane' history; 'genuine' or 'mythical' history'. 'Histoire', in *Dictionnaire de l'Académie française*, n.p.
35. Donald R. Kelley, 'Between history and system', in *Historia*, ed. Gianna Pomata and Nancy Siraisi, p.211-37 (212).

perception and story of them (the subject of Chapter 3). In the tradition of Quintilian's *Institutio oratoria*, history was considered to be the interpretative part of rhetoric, meaning that it was based on the humanizing domain of the moral, working from, in Kelley's words, 'sense rather than reason'. While it 'could never be stable and immune from error', the very idea of error, of diversion from a single, universal truth, opened up the prospect of using a narrative as an interpretative, hypothetical, experimental space in which new possibilities of narrating a story could be imagined.[36] Buffon's work combined physics and history, creating what Stéphane Schmitt calls 'une discipline tout autant explicative que descriptive' ('a discipline as explicative as it is descriptive'). Buffon intended neither to simply describe the surrounding world, nor to focus solely on one part of it such as animals or plants. Instead he chose the ambitious exercise of bringing everything together, seeking to 'subordonner autant que possible les savoirs zoologiques à une discipline bien plus vaste et ambitieuse, une histoire naturelle renouvelée et tournée vers une approche globale et explicative de la nature'. The major aspects of this approach, from the description of individual things to the discovery of general causes, are all embodied in the title of his work: *Histoire naturelle, générale et particulière, avec la description du Cabinet du roy.*[37] Moreover, Buffon sought to make the process of moving toward a broad, comprehensive search for and elucidation of the physical laws of nature a human endeavor, based upon the concurrent development of observation and reason and the ability to transform these into a language of experience.

36. Kelley, 'Between history and system', p.212, 221. On error and the development of human thought, see Wilda C. Anderson, 'Error in Buffon', *Modern language notes* 114:4 (September 1999), p.691-701.

37. '[T]o subsume zoological knowledge as much as possible into a much larger and more ambitious discipline, a renewed natural history focused on a global and explicative approach to nature'. Buffon, *Histoire naturelle*, ed. Schmitt and Crémière, 'Introduction', vol.7, p.26, 31. James Llana similarly observes: 'In assigning importance to the search for general causes, the *Encyclopédie*, like the *Histoire naturelle*, seeks to situate natural history, traditionally defined largely as an empirical, taxonomical enterprise, much closer to a rationalized, explanatory natural philosophy' ('Natural history and the *Encyclopédie*', *Journal of the history of biology* 33, 2000, p.1-25, 14).

The literary practice of natural history

The melding of disciplines and combination of discourses required an understanding as much of the status of the natural sciences as it did of letters and language. However, asking how the *Histoire naturelle* was a work of literature in the Enlightenment (and thus justifying it as an object appropriate for literary studies today) puts it in jeopardy of being read as a poetic representation of an idealized nature, little concerned with trying to grasp the real structure and operations of the world. This could not be farther from Buffon's intentions. The question of the *Histoire naturelle*'s relationship to literature is important to consider, for it provided the intellectual space and context in which to understand working between disciplines, or discourses of knowledge. Buffon must have considered how he presented and organized information, and how he in turn conceived of training his readers' powers of judgment and interpretation – activities that defined the work of the Enlightenment literary figure.

In his study of dictionary entries for 'literature' (e.g. those of Richelet and Furetière), Alain Viala has noted that at the beginning of the seventeenth century (and even continuing in the *Dictionnaire de l'Académie française* of the mid-eighteenth century) literature was defined as the erudite 'savoir de celui qui a beaucoup lu et retenu; c'est un résultat de la réception des textes, et non l'art d'en produire'.[38] It was in this area that texts of natural history were categorized, since they followed the scholarly practice of gathering and condensing knowledge from a long, received tradition of writing about nature, in both secular and religious veins. Viala shows how this philological definition of literature (whose practitioners were called 'lettrés') confronted a newer one, which designated literature as the techniques of poetic, historical and rhetorical invention. '[C]eux qui s'adonnaient à la création et à la critique littéraires', he writes, were often called 'littérateurs'.[39]

A look at articles in Diderot and D'Alembert's *Encyclopédie* reveals a desire to fuse these two definitions, or at least to consider them with respect to one another to discover what new points of view this combination could bring: the article 'Lettres' spoke of the

38. '[K]nowledge of one who has read and retained a great deal; it is the result of the reception of texts, and not the art of producing them'. Alain Viala, *La Naissance de l'écrivain: sociologie de la littérature à l'âge classique* (Paris, 1985), p.281.
39. '[T]hose who dedicated themselves to literary creation and critique'. Viala, *La Naissance de l'écrivain*, p.24-25.

importance of studying 'les sciences' (erudition, logical demonstration and systematic ordering) through the lens of letters, and particularly of the history, development and usages of language. In a cultivated society, it was argued, science and letters should support each other.[40] Jean-François Marmontel's *Encyclopédie* article 'Critique' took this idea a step further, inventing a new practice of reading and interpreting knowledge that surpassed erudition and incorporated the newer definition of literature. The practice of criticism outlined here strove to refine perception and judgment through 'un examen éclairé & un jugement équitable des productions humaines'.[41] The figure of the critic portrayed by Marmontel not only could read and organize information, but could order and present it in a way which conveyed the *rapports* between ideas and the overarching truths of the subject matter, illustrating simultaneously the content of a body of knowledge and the manner of its creation. The critic was imagined to be the guide of society, transforming erudite knowledge into useful knowledge by teaching readers how to 're-cognize' and judge information across discourses. It was thought that, with these skills, people could learn to visualize – and someday work to implement – a supposedly better, more functional world.[42]

The *Histoire naturelle* fell at this crossroads between erudition and invention, between the knowledge of letters and literary creation. On the one hand, it described the collection of the king's cabinet. In a series of plates and essays, it assembled and united a great deal of knowledge and accounts of animals, birds and minerals from all over the world.[43] Buffon's coauthor for much of the project, Louis-Jean-

40. 'Lettres', in Diderot and D'Alembert, *Encyclopédie*, vol.9, p.410. In his article 'La fiction et l'histoire naturelle au siècle des Lumières: fonctions, enjeux, dangers' (*SVEC* 2009:02, p.207-25), Benoît de Baere refers to Jaucourt's *Encyclopédie* article 'Sciences' to highlight 'l'importance des liens qui au dix-huitième siècle rattachent cette discipline [l'histoire naturelle] à la République des lettres' ('the importance of the links which connect [natural history] to the Republic of Letters', p.209).

41. '[A]n informed examination & a fair judgment of human productions'. Jean-François Marmontel, 'Critique', in Diderot and D'Alembert, *Encyclopédie*, vol.4, p.490.

42. Marmontel, 'Critique', p.493-94.

43. On the illustrations in the *Histoire naturelle*, see Elizabeth Liebman, 'Painting natures: Buffon and the art of the *Histoire naturelle*', doctoral dissertation, University of Chicago, 2003; Thierry Hoquet, *Buffon illustré: les gravures de l'*'*Histoire naturelle*' (Paris, 2007); and Buffon, *Histoire naturelle*, ed. Schmitt and Crémière, 'Introduction', vol.9 (2016), p.7-232 (42-43, 62-65).

Marie Daubenton, wrote careful, precise descriptions of quadrupeds to accompany the plates depicting the dissections and measurements of these animals.[44] Buffon collaborated with other intellectuals, especially in the later volumes on birds and minerals. The archival material from the production of these volumes (in the library of the Paris Muséum national d'histoire naturelle), along with the detailed footnotes in Buffon's articles, shows how the *Histoire naturelle* followed the erudite tradition in natural history of enumerating and classifying the sources of information about an animal – be they mythical, fabulous, traditional or empirical.[45]

In this traditional practice of natural history, according to Michel Foucault,

> to write the history of a plant or an animal was as much a matter of describing its elements or organs as of describing the resemblances that could be found in it, the virtues that it was thought to possess, the legends and stories with which it had been involved, its place in heraldry, the medicaments that were concocted from its substance, the foods it provided, what the ancients recorded of it, and what travelers might have said of it. The history of a living being was that being itself, within the whole semantic network that connected it to the world.[46]

However, Foucault's broad views of historical periods can put too much emphasis on a general definition of an epoch or intellectual practice, neglecting a subtler study of how these practices were generated, and how they were in fact discursive processes that developed and changed with the interpretation of information. That to write the history of an animal or a plant was to write everything known about it did not mean that Renaissance natural history was indiscriminate and unsystematic. Just as in Enlightenment science, certain forms of evidence were valued over others, and sources of information were subjected to specific criteria of judgment. William Ashworth famously argued that an important shift in the definition of natural history from the Renaissance to the seventeenth century was characterized by

44. On Daubenton's work in the *Histoire naturelle* see Jeff Loveland, 'Another Daubenton, another *Histoire naturelle*', *Journal of the history of biology* 39 (2006), p.457-91, and Stalnaker, *The Unfinished Enlightenment.*
45. On the footnotes and sources, especially those from Antiquity and the Renaissance, in the *Histoire naturelle* see Buffon, *Histoire naturelle*, ed. Schmitt and Crémière, 'Introduction', vol.8 (2014), p.7-131, and vol.9, p.47-81.
46. Foucault, *The Order of things*, p.129.

making 'artefactual evidence the standard for determining historical truth, [...] [ignoring] or [downplaying] the evidence of literary history'.[47] More recent studies about the history of natural history have sought to nuance Foucault's portrayal of the Renaissance *episteme* as a complete conflation of nature and culture. In particular, Brian Ogilvie writes that many Renaissance naturalists were careful to distinguish 'between the empirical study of nature and the symbolic interpretation of it'. Moreover, he argues, 'Foucault's particular interpretation excludes more widespread symbolic systems that Renaissance thinkers applied to nature. Natural theology was one: all naturalists accepted that God had created plants, animals, and minerals in six days' work.'[48] Ogilvie shows that the main purpose of this area of study 'was describing nature, cataloguing its marvelous and mundane products', reflecting the definitions of history as description discussed above.[49] As such, earlier naturalists read the world as a book whose language they needed to present in a pleasing, 'aesthetic' manner, but not interpret.[50]

However, as naturalists received changing information about the world – from the circulation and printing of previously little-known works from Antiquity, narratives and images resulting from voyages to other parts of the globe, and new ways of reading and interpreting biblical and natural history – they had to sort through these sources, deciding how to read a variety of discourses and accounts, and which ones they believed to be valid and carry authority. 'De là', Schmitt writes, 'découle une forme d'attitude critique sensiblement différente de celle qui s'appliquait aux auteurs de l'Antiquité et à leurs érudits commentateurs'.[51] By the seventeenth century, Phillip Sloan argues, a form of natural history that 'sought a causal understanding of

47. William B. Ashworth, 'Natural history and the emblematic world-view', in *Reappraisals of the Scientific Revolution*, ed. David C. Lindberg and Robert S. Westman (Cambridge, 1990), p.303-32 (322).
48. Brian Ogilvie, *The Science of describing: natural history in Renaissance Europe* (Chicago, IL, 2008), p.16.
49. Ogilvie, *The Science of Describing*, p.6.
50. According to Ogilvie, 'The Book of Nature was a powerful metaphor in the sixteenth century, as before and after, but naturalists, as naturalists, were concerned with establishing its text, not deciphering its language.' Ogilvie describes the Renaissance discipline of natural history as 'unabashedly concerned with particulars and their description [...]. If there was a prime motivation for the pursuit of natural history in the sixteenth century, it was aesthetic. Description was a way to grasp the multiplex beauty and diversity of the natural world' (*The Science of describing*, p.16, 268).
51. 'From there came a type of critical attitude visibly different from that which

natural objects in terms of historical genesis' was developed and was beginning to be considered a science in its own right. Sloan attributes this change to 'the Hermetic philosophers of the Renaissance, who were concerned to formulate a Christian philosophy on chemical premises distinct from those of Greek science and its traditions'. Authors such as Paracelsus, Fludd, Kircher and Van Helmont sought not only to describe but to explain, through reason, observations of natural phenomena. Their work constituted a 'new conception of a science which presented a historical and genetic account of the means by which the present order of things came into being under the action of natural laws'.[52]

Buffon did not reject the work of previous naturalists or the concept of the book of nature, but rather continued to pursue the changes that these naturalists had instigated in rethinking the epistemological structure of the discipline of natural history, by modifying practices of reading and interpreting. He continued to draw from a wide range of sources – direct empirical observation, works from Antiquity, Renaissance natural history and natural philosophy, travel narratives, biblical exegesis and even works of fiction such as Benoît de Maillet's *Telliamed* (1748) – but he revised the methods of reading and combining these sources. He brought a critical eye (as Marmontel would have defined 'critical') to the *Histoire naturelle* by judging and choosing between sources of information, such that his essays did not act as an accumulation of all knowledge and opinions, from definitive to conjectural, of a specific animal. By rethinking how knowledge was found and presented, Buffon changed the essence of what it meant to have certain, evident knowledge, giving new status to the study of natural history which, because it described the sublunary realm of change and corruption, had often been labeled as uncertain or conjectural. As Peter Hanns Reill argues, Buffon 'elevated history from the lowest form of understanding to a primary one'.[53] Buffon, Reill writes, transformed human engagement with nature 'into a

---

relied on authors of Antiquity and on their erudite commentators.' Buffon, *Histoire naturelle*, ed. Schmitt and Crémière, 'Introduction', vol.9, p.60.

52. Phillip R. Sloan, 'Natural history, 1670-1802', in *Companion to the history of modern science*, ed. R. C. Olby *et al.* (London, 1990), p.295-313 (297-99).

53. Peter Hanns Reill, *Vitalizing Nature in the Enlightenment* (Berkeley, CA, 2005), p.40. Contemporary debates about how the changes Buffon made to the practice of natural history were perceived, and specifically about whether and how Buffon's systems, theories and hypotheses related to the observation of reality and the discovery of its laws, are analyzed in Loveland, *Rhetoric and natural history*, ch.4.

virtue, believing that humans are better able to understand this world because they are a part of it, in fact, the most complete part because humans can reflect upon their experiences through the medium of language'.[54] Thus, in the *Histoire naturelle*, knowledge considered to be true no longer derived from logical demonstration, nor did authority derive from a scholastic chain of sources and citations going back to Antiquity.[55] These were to be discovered by the validation of the observer immersed in nature who, through the historical, temporal *and* critical process of forming analogies with the natural world, became able to think like nature and to narrate its story in written language. Reality and truth, according to Buffon, were deeply relative to their perception in thought; their discovery and presentation relied upon human interaction with the world and careful consideration and judgment of the discourses used to represent it.

Natural history taught about the world through the conversion of perception into language – eloquent, persuasive, learned language – what many Enlightenment philosophers called a language of nature. The activity of studying the world was necessarily a discursive activity, the translation of the observation and ordering of nature into words. As Chapter 1 will argue, it was the human reading and interpretation of a well-measured world, whose parts flowed together with the cadence not of a poem but of the movement of nature, from the level of the surrounding world to that of natural law. Buffon's language did not break apart the world into the smallest possible pieces (as did the taxonomic method and resulting language of Carolus Linnaeus, for example) but rather attempted to embody the universality of nature's functioning on the most general scale. As human observations, thought and language mirrored the proportions of nature on this most general scale, that of law, they became more fluent and lyrical. In other words, only through an increase in what Buffon might have called the human 'literacy' of the world, and only through increasing eloquence on the part of the explaining savant or author, could human knowledge and language express the true, most universal state of nature. The *Histoire naturelle* exemplified the Enlightenment belief in the ability of discourse to articulate the operation of the natural world.

Also see Buffon, *Histoire naturelle*, ed. Schmitt and Crémière, 'Introduction', vol.1 (2007), p.9-121 (33-42).

54. Reill, *Vitalizing nature in the Enlightenment*, p.53-54.

55. On knowledge, authority and belief in medieval thought, see Ian Hacking, *The Emergence of probability* (Cambridge, 1975), p.20-22.

By staging how to create like nature, the *Histoire naturelle* sought to prepare readers to interact with and change nature. In this way, Buffon's work came into line with a contemporary definition of fiction. In the *Encyclopédie* article 'Fiction' (linked by Diderot's editorial system of *renvois* to the article 'Critique'), Marmontel situated the role of the artist and author within the creation of invented worlds in works of fiction:

> Pour voir le monde tel qu'il est, nous n'avons qu'à le voir en lui-même; c'est un monde nouveau qu'on demande aux Arts; un monde tel qu'il devroit être, s'il n'étoit fait que pour nos plaisirs. C'est donc à l'artiste à se mettre à la place de la nature, et à disposer les choses suivant l'espece d'émotion qu'il a dessein de nous causer, comme la nature les eût disposées elle-même, si elle avoit eu pour premier objet de nous donner un spectacle riant, gracieux, ou pathétique.
>
> To see the world as it is, we have only to see it in itself; it is a new world that we ask of the Arts; a world as it should be, if it were only made for our pleasures. It is thus up to the artist to put himself in the place of nature, and to order things following the kind of emotion that he wants to arouse in us, as nature would have herself ordered them, if her first object had been to give us a joyful, graceful, or touching sight.[56]

The ultimate goal of knowledge here was its transformation by the literary mind into a model of a world more agreeable for people to inhabit.[57] In this light, the *Histoire naturelle* may be read as fictional not only because of its aesthetic value, but because it operated as a tool of knowledge, interaction and innovation. It gave the author power to invent a new order, to elicit physical, tangible change in the natural world by transmitting to nature the creative energy Buffon believed it could no longer produce itself – and making it a more pleasant, because more controlled and productive, place to live. In Buffon's work, language became not only constitutive of natural knowledge, but simultaneous with natural creation, embodying the power of

56. Marmontel, 'Fiction', in Diderot and D'Alembert, *Encyclopédie*, vol.6, p.679-80.
57. On fiction's relationship to science, Jean-François Chassay writes that 'elle aide à comprendre comment la science se forme, se réalise, et montre *comment les sciences pensent*, traduisent notre monde et infléchissent notre manière de le concevoir' ('it helps to understand how science is formed, becomes true, and shows *how the sciences think*, translate our world and inflect our manner of seeing it'; *Si la science m'était contée: des savants en littérature*, Paris, 2009, p.18; original emphasis).

nature itself. The fictional power of language furnished natural history with a moral and ethical role in the conjecturing of the story of the future relationship between humans and nature.

The 'Epoques de la nature' epitomized the culmination of a long career of understanding the study of the natural world through language. Through the voice of the natural historian, the narrative of nature was transcribed and transmitted to readers. Over the course of his oeuvre, the inventive, fictional capacities of written language developed alongside Buffon's understanding of the history of the Earth such that by the 'Epoques' their power had become literalized: language was now both a persuasive tool *and* an actual physical activity that had the ability to raise the temperature of an environment. The lyrical, dynamic quality of Buffon's writing about nature and the physical phenomenon of heat coalesced in the 'Epoques'. His language was intended to articulate to readers the activity of rewarming a cooling globe, for it represented the harmonization of mind, word and world. When the language of the text came into harmony with nature's own language of existence and creation, when the author of the text and his readers thought and spoke in the language of nature, they could then begin to change their interaction with the world, determining how to modify reality – an activity associated with Enlightenment fiction. It is this development from rational methods to physical activities, from knowing about nature to acting within it, that I will trace in this study of the language of the *Histoire naturelle.*

Summary of chapters

Buffon named the process of interaction between human thought and nature 'style'. This process regulated the relationship between knowledge and the discursive form it took, rendering these two sides of the practice of natural history inseparable in Buffon's work. As Joanna Stalnaker convincingly shows, Buffon 'made style inextricable both from the writer's individual perspective and from the world he sought to represent [...]. Thus in order to avoid the dichotomy between style and substance (and between literature and science) that has marked Buffon scholarship, we need an account of his concept of literary style that is grounded in his natural philosophy.'[58] Building on Stalnaker's analysis of the ways in which languages of history and description were produced in the *Histoire naturelle*, this book will continue to

58. Stalnaker, *The Unfinished Enlightenment*, p.34.

deepen the definition and explore the importance of language in natural knowledge, through the mechanisms of its creation and transmission – mathematics, physics, heat, hypothesis – that founded a discursive simulacrum of nature in the text.

Physical heat and discursive energy finally and directly converged in the great hybrid work of natural history and fiction, the 'Epoques de la nature', to which I will return in the last chapter of this book. In this text, the force of language was transformed from figurative to material, with very tangible consequences imagined in the real world. However, before turning to the elements of Buffon's written language of nature and how they coalesced in the 'Epoques', the manner in which he understood the relationship between perception, knowledge and language must first be examined. The study of the available, contemporary modes of invention (or what Dan Edelstein calls the 'conditions of possibility' which show how, in his words, a particular discourse 'emerged' from the intellectual circumstances of the time) is necessary to understand the poetics of the *Histoire naturelle*, and is addressed in Chapter 1.[59] What were the existing, and often competing, systems of natural philosophy used to perceive, order and write about the world, establishing the foundation by which languages of nature were understood to be accurate and authoritative? Even if he did not agree with all of their elements, it was from within these systems, and enabled by them, that Buffon was able to reinvent the study of natural history, discovering a manner in which to rearticulate the natural world that was both innovative and carried weight within the contemporary intellectual community.

In the 1740s and 1750s, Buffon used the term 'heat' in the conception and formulation of the 'Discours sur le style' to describe his understanding of natural knowledge as a language which merged nature and human thought. Chapter 2 will analyze the prose of the early *Histoire naturelle* alongside contemporary definitions of heat in mechanics and chemistry, demonstrating how Buffon's language was a crucible which merged natural and rational activity, and laying the ground for the discussion of history in the next chapter. The idea of the harmonization of world and thought was key to the production of a language of natural history, according to the poetic process of planning and writing about a subject presented in the 1753 'Discours sur le style' and analyzed in Chapter 3. The order and movement of the author's language – the energy of his words – were thought to parallel

59. Dan Edelstein, *The Enlightenment: A Genealogy* (Chicago, IL, 2010), p.17.

the order and movement of nature, translating and transmitting the operation of its laws into human language.

From Buffon's concept of style was derived that of natural history, or the story told of the events, geological and biological, of the world. In the second half of Chapter 3, I show how Buffon understood written language as a temporal process of narration: the dynamic qualities and energy of the natural world could be captured not in frozen, stagnant descriptions of the present-day world, but rather through a continuous story showing the relationships between events over a long duration of time. Buffon's words and ideas followed one another across the pages just as the cycles of nature succeeded one another across time, a model presented in his introductory essays to the first volume of the *Histoire naturelle* (1749). Readers were asked to visualize and interpret the enduring effect and power of Buffon's language as the continuous action of the laws of nature as they shaped the surface of the globe, the development of species, and the interaction between people and their environment over time.

Yet this story needed a beginning – or at least in Buffon's cosmology of continuity, a beginning was blatantly missing, especially during a time when it was considered dangerous, and perhaps even irrational, to separate the theological and natural stories of the world, and to speak of an endless universe without a moment of divine creation. His account of the origins of the world and of our solar system, discussed in Chapter 4, marked the arrival of much debate and criticism of the ideas of the *Histoire naturelle*, from theologians at the Sorbonne and contemporary philosophers and natural historians alike.[60] What interests me here is not the intersections and transgressions between natural theology and natural history in Buffon's work, but rather how a change, a rupture, an origin-event built upon the interactions between heat and history. In his 1749 'Preuves de la théorie de la terre' ('Proofs of the theory of the Earth'), Buffon revealed what was to become his controversial hypothesis for the formation of the solar system: a comet crashed into the star that was to be our Sun, breaking off pieces of this star that eventually fell into orbit around it and became the planets. The planets, including the Earth, thus began in a state of fire; at this point, the combination of the metaphorical concept of heat and energy, as the poetic process of painting vivid images with words, with the mechanical explanation of heat and energy, as

60. Roger, *Buffon*, p.251-54; Buffon, *Histoire naturelle*, ed. Schmitt and Crémière, 'Introduction', vol.1, p.107-14.

the outward movement of matter from one body to another that is perceived as change, became clearer and more central to Buffon's natural philosophy. Because the story of the comet was hypothetical, Buffon adopted a persuasive rhetoric both to convince his readers of its plausibility and to make them aware of an important principle of his work, one that would become crucial to the project of the 'Epoques de la nature': that the reality of the natural world was a construction of how people see it and the story they tell of it.

There was more to the project of the *Histoire naturelle* than simply reading and absorbing the vivacity and movement of the natural world. Buffon ultimately understood that the poetic process of the transfer of nature's energy worked as able to change not only the state of the reader's mind, but the order of nature itself. By showing what it meant for Buffon to read and rewrite the language of nature, creating a consonance between the expression of his thought in the text and the expression of the laws of natural history in its observable phenomena, the discourses of heat and history eventually changed the language of these laws. Then, through the activity of hypothesis, readers were enabled to ask whether the ultimate goal of the *Histoire naturelle* were not the harmonization of thought and the world, but rather the realization of a better world that could surpass this horizon. Through the guidance of the text, people could learn not only to understand the mechanisms of the world, but to modify them, to render them more energetic, more functional, more useful. Moreover, they could learn to appropriate the poetic force of the natural historical text, and transform it into action in the world, becoming critics and authors themselves by taking the place of Buffon's prose.

What would be the consequences of putting some of the authoritative power of nature into the hands of human beings? What happens when a text not only instructs, but joins knowledge with action, linking natural philosophy to social and historical change and striving toward larger moral and ethical aspirations? These questions are addressed in Chapter 5 of this book, which analyzes the 'Epoques de la nature' as a demonstration of the intersection between poetic and natural energy. The chapters of the 'Epoques' are structured around the gradual loss of heat over time, from the original molten state of the Earth to the present day, when Buffon argued that it was essential for people to understand how to read the history of the world in order to ensure a productive future. In this last chapter, I show how poetic energy, the way the world was represented through style, and the physical heat of nature (which Buffon claimed to have measured in a series of experiments) became one and the same. The way the language of nature is read and experienced changes how people interact with the world. In Buffon's Enlightenment dream, this would in turn give people the agency to modify nature, to author the story of its future, to create a temperate garden where there might have only been cold death.

The conclusion to this book inquires into the legacy of Buffon's work. It uses as an example the marquis de Condorcet's reading of the natural historian's contribution to natural philosophy, from his *Eloge de M. de Buffon*, the eulogy read to the Académie des sciences after Buffon's death in 1788. Leading a new generation of *savants* who were beginning to form a different definition of scientific knowledge and its uses, Condorcet rejected the effectiveness of the *Histoire naturelle* in teaching the facts of the natural world. However, he understood that the power of Buffon's language could be neither diminished nor forgotten. It might no longer illuminate the order and structure of reality, but, as a tool for inventing better worlds in the domain of fiction, Buffon's language would endure.

1. Inventing natural language: the harmonization of mind and world

> C'est de la nature que je vais écrire. Je laisserai les pensées se succéder sous ma plume, dans l'ordre même selon lequel les objets se sont offerts à ma réflexion, parce qu'elles n'en représenteront que mieux les mouvements et la marche de mon esprit.
>
> I will write about nature. I will give my thoughts leave to follow one another under my pen, in the same order as that which the objects were presented to my reflection, because they [the thoughts] will only but better represent the movements and the operation of my mind.[1]

In order to examine the inseparability of knowledge and language in the *Histoire naturelle*, and how Buffon imagined the transmission of ideas between author, reader and natural world, the question must first be addressed of how language was an epistemological tool. What did it mean to think, speak and write with order about the natural world, and in harmony with the natural world? As discussed in the introductory chapter, Buffon's practice of natural history was a combination of empirical observation, the scholarship and erudition of previous natural philosophers, and a creative, inventive power of world-making. At the base of this hybrid project was a language embodying the order and mechanism of both nature and human knowledge and thought. In this chapter, I ask what it meant to compose a language of both nature and reason that reflected their operation. To do so, I

1. Denis Diderot, 'Pensées sur l'interprétation de la nature', in *Œuvres complètes*, ed. Jean Varloot, vol.9 (Paris, 1981), p.25-111 (27).

will examine three iconic philosophical voices from the time which deeply affected the perception of nature in the Enlightenment – those of Descartes, Newton and Montesquieu – analyzing how each understood the representation of the outside world and its relationship to the human mind. Descartes's *mathesis* and Newton's vindication of optical observation of nature were combined in Montesquieu's rationalization and explanation of the study of human political history in *De l'esprit des lois*, where law was derived from both empirical observation and analogical reasoning.

Buffon's study of nature was in dialogue with these important philosophical voices because he sought to embed his work in the connection between the specificities of human reasoning and the generality of natural law. Departing from earlier understandings of natural history as the description of the surrounding world, Buffon intended not only to study each natural phenomenon in detail, but indeed, in the words of Stéphane Schmitt, to 's'élever à des considérations générales, c'est-à-dire à des lois de la nature, à des raisonnements explicatifs. Or c'est à la physique que l'on reconnaît habituellement ces ambitions.'[2] 'Physique', as discussed in the introduction, implied the search for causes and laws in the natural world. In the 'Premier discours' introducing the *Histoire naturelle*, Buffon aspired to place the discipline of natural history within the realm of physics, drawing a parallel between the order and language of ideas and the order of nature created by laws that rendered the world legible and understandable to human reason. To better understand how Buffon was able to construe the mind and the world in the same terms, this chapter will begin by examining how Descartes and then Newton interpreted the relationship between human reason and law. It will then ask how Buffon debated and incorporated these ideas into his early work. To these voices will be added that of Montesquieu, a crucial figure in the Enlightenment reinterpretation of the discipline of history. His depiction of the world as the interaction between history and law enabled, I argue, Buffon's conception of the *Histoire naturelle* as a coproduction of nature and the mind.

2. '[E]levate [his work] to general considerations, that is, to the laws of nature, to explanatory reasoning. And we usually associate these ambitions with physics.' Buffon, *Histoire naturelle*, ed. Schmitt and Crémière, 'Introduction', vol.6 (2011), p.7-88 (35).

Mathematical rules and natural laws

To speak of natural law in the early eighteenth century was to consider how much of nature could be understood by human observation and reason and in what signs or language this knowledge might be conveyed. Differing interpretations produced different discourses of nature, forming and augmenting distinctions between disciplines of study – but traditionally historians have simplified these divisions, describing eighteenth-century knowledge as being separated between mathematical and physical, Newtonian and Cartesian, ways of ordering the world. This simplification began in large part because of the way in which philosophers such as Fontenelle and Voltaire portrayed the intellectual climate of their own time.[3] However, what the two categories, Newtonian and Cartesian, designated was quite complex and overlapped in interesting ways. Much of the initial reception of Newton's *Principia mathematica* in France was filtered through the lenses of Cartesian philosophy.[4] In particular, Newton's mathematical writings were interpreted by French intellectuals through Descartes's concept of relationships between ideas and things as the foundation of both true scientific knowledge and the language of this truth.[5]

Cartesian mathematical thought was deeply tied to interpretations of Newtonian natural law in the early French Enlightenment. The ordering, proportioning and measuring of abstract ideas allowed for the translation of thought into a language. This language was the key to the interpretation of the outside world; it rendered thought commensurable with nature, allowing observers to study and express the operation of its laws. Natural law, at the intersection of the mind and the world, was construed as a discursive entity. It epitomized

3. Pierre Brunet emphasizes this divide in *L'Introduction des théories de Newton en France au XVIIIe siècle* (Paris, 1931). Revised analyses of the reception of Newton's work in eighteenth-century France include Peter Dear, *Revolutionizing the sciences: European knowledge and its ambitions, 1500-1700* (Princeton, NJ, 2001); Henry Guerlac, *Newton on the Continent* (Ithaca, NY, 1981); J. B. Shank, *The Newton wars and the beginning of the French Enlightenment* (Chicago, IL, 2008); Mary Terrall, *The Man who flattened the Earth: Maupertuis and the sciences in the Enlightenment* (Chicago, IL, 2006).
4. See J. B. Shank, '"There was no such thing as the 'Newtonian Revolution', and the French initiated it": eighteenth-century mechanics in France before Maupertuis', *Early science and medicine* 9:3 (2004), p.257-92 (262-63).
5. See Guerlac, *Newton on the Continent*, p.58-59; and Jean Ehrard, *L'Idée de nature en France dans la première moitié du XVIIIe siècle*, 2 vols (Paris, 1963), vol.1, p.61-91.

the space in which natural philosophers claimed to speak the true language of nature: not simply a representation of the world, but tantamount to the truest form of knowledge, representing the act of bringing the order and movement of their thoughts into harmony with the order and movement of the world.

Because he argued that the world always behaved in the same way through divine laws, Newton's natural philosophy legitimized empirical, and especially visual, observation. However, this vindication of observation was treated cautiously by French Enlightenment philosophers, because of the deep influence of Cartesian thought, which taught to mistrust the senses, separating them from the reasoning mind. The contents of the *Histoire naturelle* relied heavily on visual observation, but the process through which these observations were generalized into laws reflected a unique combination of Newtonian and Cartesian thought. Buffon used mathematics both as a way to universalize observations and to make nature legible through human reason; from these two viewpoints his language of nature was born.

I begin with an analysis of Descartes, for it is upon this basis that I believe Buffon, like many other French intellectuals at the time, read and incorporated Newton's thought. In order to appreciate Newton's arguments on the depiction of nature through law, it is first necessary to address the depiction of thought through the rules, or laws, of reason: the order and disposition of ideas in the mind, or what Descartes named *mathesis universalis*. More than a subject matter, mathematics in the seventeenth and eighteenth centuries consisted of the rules for ordering a set of ideas and expressing these ideas in a language that embodied this order.[6] Order, in turn, was the indication of the relationships between ideas in a chain of reasoning, as well as a way to conceptualize the relationship between the parts and the whole. Mathematics was a discursive tool applicable not only to quantities but to the quality of the relationships between all manner of ideas.[7] In the

6. For a history of the transition of mathematics from the seventeenth and eighteenth centuries, as an activity of making relationships in the mind, and translating these into language, to its incarnation in the nineteenth century as an innate part of the natural world, see Jean Dieudonné, *Pour l'honneur de l'esprit humain: les mathématiques aujourd'hui* (Paris, 1987), especially Chapters 2-4.
7. René Descartes, *Règles pour la direction de l'esprit*, ed. and translated by Jacques Brunschwig (Paris, 1997), p.96. English translations are taken from René Descartes, *Rules for the direction of the mind*, translated by Laurence J. Lafleur (Indianapolis, IN, 1961).

Règles pour la direction de l'esprit Descartes introduced *mathesis universalis*, defining mathematics not as an autonomous discipline but rather as an activity of the reasoning mind, which created ordered and valid knowledge, the definition of a science.[8] It was a method that trained the mind to understand, compare and order ideas, no matter the area of study: 'c'est l'ordre et la mesure que l'on examine, [...] peu importe que cette mesure soit à chercher dans des nombres, des figures, des astres, des sons, ou quelque autre objet.'[9]

'Measure' referred to the way in which ideas were arranged with respect to one another and to an argument as a whole. Within a group of ideas, Descartes explained, one must 'identifier celui qui est le plus simple, et voir comment tous les autres en sont, soit plus, soit moins, soit également éloignés.'[10] To measure was to determine how one idea could be known from a previous one. This was the first step in the act of reasoning, or finding the ratio or relationship between ideas in order to perceive the more general operation of an argument. Measurement, and hence reason, were forms of comparison. They allowed for the discovery of what Descartes called 'the absolute', distinguishing this from 'the relative'. The former, the absolute, encompassed the essence of a series of ideas, 'tout ce qui contient en soi, à l'état pur et simple, la nature sur laquelle porte la question'.[11] The latter, the relative, represented instantiations or degrees of the

8. The *Rules* were begun around 1619 and published after Descartes's death. Descartes left the *Rules* incomplete, abandoning them to create a more universal method for finding the truth. In their unfinished state, the *Rules* did not hold an authoritative, definitive position in the body of his works and it is unclear how Descartes considered them with respect to his other texts. However, as the work in which Descartes discussed *mathesis universalis*, the *Rules* had an important influence on many seventeenth- and eighteenth-century authors. See Kim Sang Ong-Van-Cung, 'Préface', in Descartes, *Règles*, p.7-9.
9. '[L]a science générale qui explique tout ce qu'il est possible de rechercher touchant l'ordre et la mesure', Descartes, *Règles*, p.96. In translation, mathematics is a 'general science which explains everything which can be asked about order and measure.' Descartes writes that, when making a science, 'order and measure are considered, [...] and it makes no difference whether such measure is sought in numbers, or figures, or stars, or sounds, or any other object whatever', Descartes, *Rules*, p.17.
10. Descartes, *Règles*, p.99; 'which one is the simplest of all, and how far from this one each of the others is removed, either more, or less, or equally', Descartes, *Rules*, p.20.
11. Descartes, *Règles*, p.100; 'whatever contains in itself the pure and simple essence with which we are concerned', Descartes, *Rules*, p.20.

absolute, local analogies between one idea and the next: 'ce qui est de la même nature, ou du moins y tient par un côté par où l'on peut la rattacher à l'absolu, et l'en déduire'.[12] The relative could be compared or measured with respect to the essential, absolute nature of the question, determining the extent to which it depended upon or deviated from this absolute. The better the parts, or relative nodes, of an argument worked together, the clearer, simpler and more true its absolute meaning could become.

The chain of ideas formed from this comparative act of gauging the distance between relative and absolute revealed the distinct yet analogous relationships between adjoining ideas: their 'connections with one another' and their 'natural order' or disposition with respect to the absolute nature of the series itself.[13] Each idea was embodied by 'un signe unique' ('a unique [sign]'); each sign or unit symbolized a step or degree of the reasoning process.[14] What made a set of ideas 'natural' was the discovery of a single rule or standard of measurement by which to order and explain them. Each element of the set was understood as a 'unit', in proportion with and dependent upon the essence of the whole.[15] A series had natural order and reason when all of its parts corresponded or were analogous to the absolute, defining principle of the order.[16] The language of the whole then could be used to speak of all its constituent parts. Moreover, a set of ideas was natural if it did not surpass the limit and capacity of human intellect, of what was considered knowable. Ideas and the things they represented were

12. Descartes, *Règles*, p.100; 'that which, while having the same nature or at least participating in it to some degree, is secondary in that it can be traced back to the absolute, and deduced from it by some chain of reasoning', Descartes, *Rules*, p.20-21.
13. Descartes, *Rules*, p.21; 'connexion mutuelle', 'ordre naturel', Descartes, *Règles*, p.100-101.
14. Descartes, *Règles*, p.180; Descartes, *Rules*, p.81. In this English translation, the word 'signe' is given as 'symbol'. I do not agree with this, and have replaced it with 'sign' in the above text.
15. Descartes, *Rules*, p.72; *Règles*, p.178-79.
16. In *La Logique, ou l'Art de penser* (1662), Arnauld and Nicole defined reason as the arrangement of ideas, determining either 'de combien l'une surpasse l'autre' ('by how much one surpasses the other') or 'de quelle manière l'une est contenue dans l'autre' ('in what way one is contained in the other'). The authors defined proportion between ideas as 'l'égalité de l'une ou de l'autre de ces sortes d'habitudes' ('the equivalence of one or other of these practices [of measurement]'). See Antoine Arnauld and Pierre Nicole, *La Logique, ou l'Art de penser*, ed. Charles Jourdain (Paris, 1992), p.293.

within reason or measurable, Descartes wrote, 'dans la mesure où elles sont accessibles à l'entendement'.[17]

Through Cartesian eyes, the operation of the outside, natural world could only be elucidated through a reinvention of the world according to the internally rigorous rules of reason. Science, or ordered knowledge, was a product of the inner light of the spirit. Isaac Newton also believed that to think mathematically was to think naturally, but for him nature was the creation not of the mind but directly of God. Because of this he famously believed that reasoning through mathematics could explain why natural phenomena occurred, only how they happened.

Newton explained this philosophy in a short work, the *General scholium*, an appendix to the third book, entitled the *System of the world*, of the *Principia mathematica*.[18] Here Newton designated God as the final cause and explanation for all phenomena in the universe. He argued that people do not possess a complete understanding or idea of God, and can only perceive a small sliver of all of God's universe and power. Newton's God was not absolute in a Cartesian sense; he was described as outside of the relativity of human thought. Newton stated that 'he exists always and everywhere'.[19] This 'Supreme God' or 'Being eternal, infinite, absolutely perfect' was not analogous or comprehensible to the perception of the human senses. The approximate, imperfect language produced by observation was incapable of describing or making sense of God who, Newton argued, existed

> in a manner not at all human, in a manner not at all corporeal, in a manner utterly unknown to us [...]. He is utterly void of all body and bodily figure, and can therefore neither be seen, nor heard, nor touched [...]. We have ideas of his attributes, but what the real substance of anything is we know not.[20]

Nature, the essence of God's creation, was absolute because it subsisted independently of human thought and history. The existence and

17. Descartes, *Règles*, p.118; 'in so far as [they] can be dealt with by the intellect', Descartes, *Rules*, p.34.
18. The *General scholium* was first published in the second edition of the *Principia*, in 1713, and some changes were made to it in the third edition of 1726. See Florian Cajori's appendix to Newton's *Mathematical principles of natural philosophy and his system of the world*, translated by Andrew Motte, ed. Florian Cajori (Berkeley, CA, 1960), p.668-70 (n.52).
19. Newton, *Mathematical principles*, p.545.
20. Newton, *Mathematical principles*, p.545-46.

presence of Newton's God was suggested 'from the appearance of things', that is, from the order of observed phenomena. People were limited, Newton thought, to knowing God in this dimension, through surrounding physical objects and events, as expressed through language derived from the senses. They could understand or speak of God only in terms relative to themselves: in Newton's words, 'from the ways of mankind by a certain similitude'.[21]

Since people could not truly know Newton's God from within the limits or measure of their intelligence, the *General scholium* implied that they must accept God's absolute power, eternity and infinity through faith, as through a necessary subservience. Newton wrote that as the author, judge and legislator of the universe God 'has dominion' over everything. God structured the world through the giving of laws. The idea of legality ultimately was derived not from the process of reason but from the sovereignty of the law-maker himself. As such, people had to obey and follow God's laws as 'his servants' who 'adore him'; in turn, giving and obeying laws defined God's 'dominion'.[22]

With respect to the abstract totality of the universe, Newton considered human perception to be limited and inconsistent because always and necessarily practiced in a relative manner. When distinguishing between absolute and relative motion, for instance, he explained that objects seemed to human eyes to move faster or slower in different moments and conditions, appearing to change in the process. However, these appearances existed relative to people's limited observational capacities, which in fact were the only aspect of this experience that underwent change in the form of imperfection or historical contingency.[23] By contrast, nature always behaved in the same manner, irrespective of local historical changes and errors, in obedience and accordance with law. Newton's universe was subject to what Frank Manuel has called the same eternal 'system of interpretation' and measurement.[24] The world, Newton thought, was only

21. Newton, *Mathematical principles*, p.546.
22. Newton, *Mathematical principles*, p.544-46. On the role of religion in Newton's work, see Frank Manuel, *The Religion of Isaac Newton: the Freemantle Lectures, 1973* (Oxford, 1974); and Richard S. Westfall, 'The rise of science and the decline of Orthodox Christianity: a study of Kepler, Descartes, and Newton', in *God and nature: historical essays on the encounter between Christianity and science*, ed. David C. Lindberg and Ronald L. Numbers (Berkeley, CA, 1986), p.218-37.
23. See Newton, *Mathematical principles*, p.6, 12.
24. Manuel, *The Religion of Isaac Newton*, p.91.

really knowable because it did not undergo historical change. All its parts remained identical to themselves over time and space.

However, it was this aspect of Newton's philosophy that allowed nature – or at least the small part of it accessible to human senses – to be known and represented in the language that most closely approached perfection, that of natural law, written in mathematical terms. Laws established the conditions that held such that observations of the same phenomena could be consistently repeated, verified and epitomized into formulas. The trajectory of nature might have already been decided by God, but as long as people obeyed God by observing the expected and regular conditions of laws, a stable and reliable science and language of nature could be built. It was in this sense that God, and consequently God's laws, became the ubiquitous fixed point from which every motion in the universe might be measured at any moment in time, and then expressed in the mathematical terms of relationship that appealed to human reason and moreover, Newton thought, approached as closely as possible the true operation of nature. Even though God's choices for making the world as it was could not be explained, this interpretation of law allowed observation of the world through the senses to become a dependable way to generate knowledge.

Newton's thought represents a transitional moment in the relationship between natural law and the ability of human reason to understand and represent law. This shifting perspective hinged on the definition of the term 'natural' which, for Newton, was that which had been ordained by God. According to this way of thinking, a universal and ubiquitous set of laws, of which humans were the obedient subjects, already had been written by God into the nature of the world. The entirety of the divine plan could not be reasonably known and essentially had to be accepted on the basis of faith.

Alongside Descartes's argument that the natural was the creation of human reason, seventeenth-century theorists such as Pufendorf, Grotius and Leibniz, however, began to ask how political customs and practices – the products of the unpredictable domains of human culture and history – could be explained with respect to God's design. They distinguished this kind of law, which they called 'natural', from divine law by systematizing the former, making it legible to reason.[25] Additionally, such thinkers raised questions about defining and studying the causes and mechanics of the common, natural course

25. Donald R. Kelley, 'Law', in *The Cambridge history of political thought, 1450-1700*, ed. J. H. Burns and Mark Goldie (Cambridge, 1991), p.66-94 (88, 92).

of things. In particular, they debated over the origins and sustainability of motion in the physical world, asking whether God had set the universe in motion at the beginning, and thus allowing movement to be studied according to the same, fixed laws, or whether God continuously intervened, deciding at each instant the trajectory of objects and consequently the course of history.[26] As the perceived capacities and boundaries of human reason changed, the idea of judgment gained importance with respect to divine authority and became a key factor in determining the 'natural'. It was possible to imagine what T. J. Hochstrasser has called 'an increasingly intricate framework of comparisons [...] between the rational truths accessible to man and what was alleged could be known of the mind of God'.[27] Relativism, through comparison and judgment of the human and divine realms of law, was injected into both Cartesian and Newtonian philosophy. Because it was meant to apply to every time and place universally, Newton's conception of natural law granted permission to study both the physical and the moral, the natural and the historical, as logical systems which shared the same method and language. Buffon inherited a hybridized way of thinking in which politics, religion, economy and nature could be comprehended according to the same idea of natural law, one which stemmed from the rational ability to order and define the surrounding world.

In Buffon's work, two types of nature confronted one another: the nature of reason, 'natural' because based upon an internal rigorous system of measurement, and the nature of the outside world, ordered by its own rules, those of law. For Buffon, truth was not generated by the discovery of the simplest and most absolute ratio of ideas in the mind, which had been the defining principle of an entire science for Descartes. He worked with respect to the outside world, and thus his observations of it were defined according to the laws of nature. However, Cartesian concepts of measurement and proportion remained crucial to comprehending his method of studying and composing natural history. The capacity of reason to grasp and articulate the order of the outside world made it possible to be less obedient to an unknown divine plan (as Newton had demanded in the *General scholium*) and allowed for a more creative, inventive power to represent nature in thought and language. What Descartes intended

26. Ehrard, *L'Idée de nature en France*, vol.1, p.81-84.
27. T. J. Hochstrasser, *Natural law theories in the early Enlightenment* (Cambridge, 2000), p.4-5.

as a transparent, rational and natural process of translation from thought to language was applied to the study of the nature of the physical world. The language of the *Histoire naturelle* sought to find the relationships between ideas of the world and the world itself, and to bring these two into proportion with one another.

Buffon and natural law: relativizing perception

How did Buffon use this combined definition of reason and law to establish the relationship between perception, knowledge and language in the *Histoire naturelle*? Answering this question necessitates returning to the beginning of his career to understand how he assimilated and combined Cartesian and Newtonian thought. In the Académie royale des sciences, where he began his career, Buffon did not start in a discipline related to natural history, but rather in the area of mathematics. His early career was marked by great aspirations to mathematical study. In fact, it was political developments within the Académie des sciences which necessitated his transfer to the role of botanist. His experience at the intersection of mathematical and physical knowledge may help to explain the manner in which he represented and communicated natural knowledge in his work.

In 1734, Buffon was elected to the Académie royale des sciences in the rank of *adjoint mécanicien*.[28] At the time, mechanics was the branch of mathematics concerned with laws of motion and forces.[29] The beginning of his career was thus shaped by a field which might appear quite foreign and removed from the study of animals, birds and minerals for which he later became known. However, when understood in the Cartesian sense, not as the science of numbers and quantities, but as a method of ordering and comprehending the world

28. Réaumur, famous for his studies of insects, was also grouped under 'mécanicien' in the Académie royale des sciences. This may have also been related to the ambiguous nature of natural history at the time, the study of nature falling between disciplines of anatomy, botany and chemistry, as well as between methods of description and the search for physical causes and laws. See Buffon, *Histoire naturelle*, ed. Schmitt and Crémière, 'Introduction', p.18-31.
29. Roger Hahn explains that areas of knowledge in the Académie des sciences were, until 1785 and Lavoisier's reforms, distinguished according to two general categories: 'the "mathematical" (exact) sciences and the "physical" (natural) sciences. In the former category were geometry (mathematics), astronomy, and mechanics; in the latter, anatomy, chemistry, and botany' (*The Anatomy of a scientific institution*, p.98-99).

through figures of relation, mathematics can be said to have deeply impacted Buffon's reading of the natural world in the *Histoire naturelle.* From his education and his correspondence at the time, it appears that Buffon initially was interested in a career in mathematics.[30] Jacques Roger mentions some of Buffon's Parisian allies in the 1730s, notably two well-known mathematicians, Alexis Claude Clairaut and Pierre Louis Moreau de Maupertuis.[31] Buffon was particularly intrigued by the study of probabilities, and was reading the major works on this subject; in a speech given in early 1733 to the Académie des sciences, entitled *Mémoire sur le jeu de franc-carreau* (*Memoir on the game of franc-carreau*), he explored the meaning and use of games in the growing science of probability.[32]

In the Académie royale des sciences, Buffon was subsequently moved to a position in botany, in the class of physics. While it was common for members of the Academy to move within the areas of mathematics or physics, changing from one class to the other was rare. When it happened, it was usually a practical matter, resulting not from the member's intellectual interests but from what Roger Hahn calls the desire 'to advance a scientist toward pension status'. Buffon, according to Hahn, was one of the major exceptions to this practice: his switch appears to have occurred because of an intellectual rift with some of his *mécanicien* colleagues, revealing how Buffon was already defining and using mathematics very differently from many of his colleagues.[33]

30. See, for example, his correspondence with Gabriel Cramer in Françoise Weil, 'La correspondance Buffon–Cramer', *Revue d'histoire des sciences et de leurs applications* 14:2 (1961), p.97-136.
31. Roger, *Buffon*, p.30, 37. For instance, in a 1732 letter to the Swiss mathematician Gabriel Cramer, Clairaut mentioned that Buffon had already sent him 'une méthode de trouver les centres de gravité fort ingénieuse' ('a very ingenious method of finding centers of gravity'). Quoted in Lesley Hanks, *Buffon avant l'*'*Histoire naturelle*' (Paris, 1966), p.38.
32. Although this text has not yet been found, what Buffon republished in his 'Essais d'arithmétique morale' in the fourth *Supplément* to the *Histoire naturelle* (1777) is believed to be quite close to the missing work, as these 'Essais' were composed from earlier *mémoires* that Buffon read to the Académie des sciences in the 1730s and 1740s. See Roger, *Buffon*, p.37.
33. Hahn, *The Anatomy of a scientific institution*, p.99. Lesley Hanks argues that Buffon thought that 'la plupart des propriétés physiques ne se laissent pas mathématiser. Il préconise donc le retour à l'expérience directe, et, bien qu'il reconnaisse aux mathématiques une certaine applicabilité, il leur refuse toute valeur comme instrument de découverte' ('most physical properties cannot be mathematized. He thus encouraged a return to direct experience, and although he still thought

Buffon's move to natural history has been labeled as a methodological change in his thought, from abstract reasoning to acquiring knowledge through the physical senses.[34] However, such an interpretation does not fully take into account the Cartesian idea of mathematics as a method of reasoning, crucial to the process of ordering thoughts about the world and conveying them in written language. Indeed, Buffon's early studies in mathematics were central to the construction of his science of natural history. He did not reject mathematics for the study of the physical world, but rather tempered and combined Newtonian and Cartesian interpretations. The beginning of his career appears to have been marked by a careful consideration of the purpose and place of mathematics in the examination of the material world. When he critiqued mathematics in the introductory essay to the *Histoire naturelle*, it was not to refute an entire science, but rather to comment on the application of mathematics to knowledge of the world, that is, on the search for the best method of reading nature and discerning its laws.

In this critique of the method of mathematical reasoning in the 'Premier discours' of the *Histoire naturelle* Buffon asked readers to question the origin and truth of ideas about the world. From the Cartesian assumption that primary ideas in the mind are essentially true, he argued, the mathematician builds a science. He described the results of this science, however, as 'abstract', lacking relationships to actual things in the world. There was no natural law or physical necessity guiding their creation: the mind alone produced the basic principles and ensuing results, such that every aspect of this chain of reasoning became a circular reflection, a function, of its starting point. 'Nous avons fait les suppositions, nous les avons combinées de toutes les façons, ce corps de combinaisons est la science mathématique; il n'y a donc rien dans cette science que ce que nous y avons mis [...]; ainsi les vérités mathématiques ne sont que les répétitions exactes des définitions ou suppositions.' Buffon thought that all conclusions drawn

---

mathematics had some applicability, he denied them all value as an instrument of discovery'; Hanks, *Buffon avant l'"Histoire naturelle'*, p.26).

34. A change which Jacques Roger calls Buffon's 'refus de l'absolu' ('refusal of the absolute'), and which Hoquet describes as a switch from the demonstrative truth of mathematics to the probabilistic certainty of physics. See Roger, *Buffon*, p.60-65; Hoquet, *Buffon*, p.321. For a detailed study of the mathematical ideas which influenced Buffon early in his career, as well as an analysis of how Buffon used mathematics in lesser-known texts such as the 'Essais d'arithmétique morale', see Loveland, *Rhetoric and natural history*, ch.5.

from this method are nothing but 'des expressions différentes de la même chose' ('different expressions of the same thing').[35]

One outcome of this kind of reasoning, according to Buffon, was that mathematical truths never varied in degree: once the first principles had been established as true, every concept derived from them carried the same kind and amount of truth.[36] As these truths were all of the same nature, '[c]e qu'on appelle vérités mathématiques se réduit donc à des identités d'idées & n'a aucune réalité' ('[w]hat we call mathematical truths thus is reduced to the identities of ideas and has no reality').[37] Mathematical truths in themselves said nothing about the status or measurement of phenomena in the outside world, or about the perspective of the viewer. If mathematical truths were abstract, or relative only to themselves, then a standard of measurement between them and the truths of the outside world was lacking, as was an ability to translate, to create a language, between the two domains.

This problem of translation was proof to Buffon that mathematical quantities were not the true language of nature. In a debate that he had with his *mécanicien* colleague Alexis Clairaut in 1745 at the Académie des sciences, Buffon asked how the language of numbers, the signs representing quantities, could be used to speak of things in the physical universe:

> [E]xiste-il dans la Nature un coëfficient comme un 4, un 5, un 6, &c? & n'y a-t-il pas de l'absurdité à supposer qu'un nombre puisse exister réellement, ou qu'un coëfficient puisse être une qualité essentielle à la matière? il faudroit pour cela qu'il y eût dans la Nature des phénomènes purement numériques […]: il faut donc que M. Clairaut commence par nous prouver que les nombres sont des êtres réels actuellement existans dans la Nature, ou que les coëfficiens sont des qualités physiques, s'il veut que nous convenions avec lui que la loi de l'attraction ou toute autre loi physique, puisse être exprimée par deux ou plusieurs termes [mathématiques].
>
> [D]oes there exist in Nature a coefficient like a 4, a 5, a 6, etc.? and is it not absurd to suppose that a number could really exist, or that a

35. 'We have made suppositions, we have combined them in every way, this body of combinations is the science of mathematics; there is thus nothing in this science but that which we have placed there […]; thus mathematical truths are nothing but the exact repetitions of definitions or suppositions.' Buffon, *Histoire naturelle* (1749-1767), vol.1, p.53-54.
36. Buffon, *Histoire naturelle* (1749-1767), vol.1, p.56-57.
37. Buffon, *Histoire naturelle* (1749-1767), vol.1, p.24.

> coefficient could be an essential quality of matter? For this, it would be necessary to have purely numeric phenomena in Nature [...]: Mr Clairaut must therefore begin by proving to us that numbers are real beings actually existing in Nature, or that coefficients are physical qualities, if he wants us to agree with him that the law of attraction or any other physical law could be expressed by two or several [mathematical] terms.[38]

Numbers and the equations formed by them could not be assumed to embody the language of the universe. By asking if 'a coefficient could be an essential quality of matter', Buffon accused his colleague of imposing an arbitrary language upon the world and thus of revealing nothing but the nature of his own thought.

A universal, natural language was, for Buffon, based on a very different, relativistic understanding of mathematics. His debate with Clairaut focused on how to express Newton's law of gravitation in the mathematical language of a formula. Unlike Clairaut, who argued for the increasing complexification of this formula to fit specific observations of the Moon's movement, Buffon held that the simplest expression of Newton's inverse square law (the principle that gravitational force is inversely proportional to the square of the distance between two objects) was the most accurate representation of this natural phenomenon. For him, the defining characteristic of Newton's law of gravitation was the generality of its scope and expression, representing motion in the world in a single explanation.[39] He believed that the truth of Newton's conclusions about the movement of the planets should be upheld, all the more so because they were confirmed by the observations and conclusions of other members of the astronomical community. Buffon agreed with Newton's list of 'Rules' for studying natural philosophy in the *Principia*, taking them as the philosophical foundation for the study of the world. He seemed especially convinced by the first and second rules, where Newton stated that '[w]e are to admit no more causes of natural things than such as are both true and sufficient to explain their appearances', and that '[t]herefore to the same natural effects we must, as far as possible, assign the same causes'.[40]

38. Buffon, 'Seconde addition au mémoire qui a pour titre: "Réflexions sur la loi de l'attraction"', in *Histoire et mémoires de l'Académie royale des sciences* (Paris, Imprimerie royale, 1749), p.580-86 (582).
39. Buffon, 'Réflexions sur la loi de l'attraction', in *Histoire et mémoires de l'Académie royale des sciences* (Paris, Imprimerie royale, 1749), p.493-500 (494-96).
40. Newton, *Mathematical principles*, p.398.

Like Newton, Buffon understood nature as amenable to study on a general level because all of its components were manifestations of the same laws. Observed phenomena thus could be interpreted according to these laws, using the same relational set of ideas, from which would be developed a shared, universal language – the Cartesian *mathesis universalis*. Near the end of his debate with Clairaut, he concluded that 'une loi en Physique n'est loi que parce que sa mesure est simple' ('a law in Physics is only a law because its measurement is simple'). Because it was 'simple', able to be represented as a single mathematical term, Buffon understood the inverse square law to be universal and all-encompassing, bringing the effects underneath it into analogous relationship with one another (even if it did not bend and become more complex to fit all observed phenomena, as Clairaut tried to do in his formulation of the law).[41] Through observation, reality, according to Buffon, should appear increasingly to be structured by the single essence of law. He argued in particular against Clairaut's complicated, long formula for expressing the movement of the Moon, claiming that this tendency to believe that, by changing the rules of a specific kind of motion, one could modify an overall law of nature was a misjudgment made on a limited (and not a general) scale. Buffon accused Clairaut of trying to grasp all of nature through a narrow vision that was not in proportion to nature itself, and 'cette supposition [...] non seulement anéantiroit la loi de l'attraction [...], mais même donneroit entrée à toutes les loix possibles & imaginables', an unending series of approximations in the place of one mathematically simple and absolute law.[42]

Buffon argued that the truths obtained from the observation of the outside world, which he called 'les vérités physiques', 'ne sont nullement arbitraires & ne dépendent point de nous' ('are in no way arbitrary and do not depend upon us'). Combining Cartesian and Newtonian thought, Buffon held that whether or not people trusted in their observations of the physical world did not detract from the reality that these phenomena were determined by invariable, natural laws. In other words, the essence of natural phenomena had little to do with human reason or systems. Yet, built into the act of observing nature was the possibility to learn, to interact with and to slowly

41. Buffon, 'Réflexions sur la loi de l'attraction', p.497.

42. '[T]his supposition [...] not only destroys the law of attraction [...], but even gives way to all possible and imaginable laws.' Buffon, 'Réflexions sur la loi de l'attraction', p.497.

merge rational and physical worlds. In his 'Essais d'arithmétique morale', Buffon wrote that '[l]'évidence qui nous vient par les sens a des rapports que l'on doit comparer et dont on peut estimer la mesure' ('[t]he evidence that comes to us via the senses has relationships that one can compare and of which one can evaluate the measure').[43] This kind of evidence possessed 'rapports', like hooks which could, through comparison, be used to link observations, strengthening both their probability and their argumentative power, making them more believable.

What Buffon called 'facts' or 'physical certainties' could be understood as a combination of an event's mathematical probability of happening with the observer's repeated experience of it: the more these two activities coincided, the more believable and certain the event became, with a probabilistic sum so immense that it 'nous force à croire' ('forces us to believe').[44] The comparison of the calculation of probability with the degree of human belief in this calculation became the standard of measure by which knowledge of the world was made. Precision was not about a specific number or result, but rather had degrees. A calculation became more precise as it acquired more possibilities of comparison – more of 'la force des analogies' ('the force of analogies') – with other existing calculations of the same phenomena. Buffon declared to Clairaut that 'jusqu'ici tout s'accorde' ('up until now, everything is in agreement') in terms of observations of the universe and their corresponding laws.[45] Probability and certainty were not the same thing as abstract truths, but in the study of the natural world Buffon allowed them to be treated in the same way, as 'une vérité pour nous' ('a truth for us').[46] Facts were the result of a dialogue of measurement and comparison between humans and nature. Knowledge of nature in the *Histoire naturelle* occupied

43. Buffon, 'Essais d'arithmétique morale', in *Histoire naturelle, générale et particulière: supplément*, vol.4 (Paris, Imprimerie royale, 1777), p.46-109 (47-48). This work is reprinted in *Un Autre Buffon*, ed. Jacques Roger (Paris, 1977), p.32-91.
44. Buffon, 'Essais d'arithmétique morale', p.52. On Buffon's relativistic and probabilistic thought see Thierry Martin, 'Certitude et probabilité selon Buffon', *L'Enseignement philosophique* 49:3 (January-February 1999), p.5-18; and Phillip R. Sloan, 'Kant on the history of nature: the ambiguous heritage of the critical philosophy for natural history', *Studies in history and philosophy of science, part C: studies in history and philosophy of biological and biomedical sciences* 37:4 (December 2006), p.627-48.
45. Buffon, 'Réflexions sur la loi de l'attraction', p.493-96.
46. Buffon, *Histoire naturelle* (1749-1767), vol.1, p.57.

an interactive space between the activity of human reason and the phenomena of the outside world.

Like his other contemporaries working in the domain of natural philosophy, Buffon did not blindly agree with either Descartes or Newton. He worked strategically with their philosophies, manipulating ideas of measurement, rule and law to produce the model of nature of the *Histoire naturelle* in which knowledge about the world came from the harmonious collection of 'analogies' between both similar phenomena and similar interpretations of these phenomena. In the *Histoire naturelle*, unlike in Newton's *General scholium*, there was no limit to human observational and perceptive abilities and no point of abstraction, no limit of expression in the face of divine power, after which they could go no further. In his introductory essay Buffon wrote that 'tous les faits de la Nature que nous avons observez, ou que nous pourrons observer, sont autant de vérités, ainsi nous pouvons en augmenter le nombre autant qu'il nous plaira, en multipliant nos observations; notre science n'est ici bornée que par les limites de l'Univers.'[47] The relationship between people and nature in the *Histoire naturelle* was not a static one: the power and scope of natural law continuously evolved as people studied it, becoming not only more general but more amenable to human thought and ingenuity. Mathematics and natural law were not only structuring principles of knowledge, but indeed deeply discursive, inventive strategies in the *Histoire naturelle*. These ideas of intervention, innovation and reasoned modification of nature seem to be derived from Montesquieu's philosophy of history as a product of the interaction of reason and law.

Inventing and intervening: Montesquieu and the natural laws of history

Harmonizing mind and nature required understanding both measurement and also natural law. Newton's conception of law permitted Buffon to approach his study of nature as a unified and universal whole which operated everywhere according to the same laws. He could assume that the world worked as a system in which all parts, manifestations of the overarching law, existed in proportion

47. '[A]ll the facts of Nature that we have observed, or that we can observe, are so many truths, and thus we can increase the number of truths as much as it pleases us, by multiplying our observations; our knowledge is only bounded here by the limits of the Universe.' Buffon, *Histoire naturelle* (1749-1767), vol.1, p.57.

with respect to one another. The search for the reason of the world, the units or language through which it could be measured, proportioned and thus comprehended, was validated by Newton's philosophy and inductive method, and could be accomplished by discovering the relationships between phenomena, and then articulating these specific relationships in the more general terms of cause and effect. Yet, unlike Newton, for Buffon, perceiving nature was not simply about obeying a divinely determined structure, but also about discovering the language to best able to capture this order, and eventually innovating upon its expressive ability, searching for a better, more useful manner of observing, writing about and interacting with nature. It was from Montesquieu's interpretation of law that Buffon seems to have borrowed this idea of observation as an interactive process.

Although he is sometimes thought of as the Newton of political history, Montesquieu naturalized the world in a very different manner from Newton. While he did believe that the science of history which he studied had inherent laws that produced and ordered it, Montesquieu showed that it could and should be shaped by human perception of and interaction with these laws. History, in this way, had two authors, a combination both of what its nature did, through its laws, and of what people did, following and interpreting these laws. In *De l'esprit des lois* (1748), Montesquieu read historical events not simply as the result of higher causes or laws (ultimately given by God), but in terms of human activity, or how people manipulated laws to produce new events and even new versions of law.

If Newton described the eternal laws, the proof of God's dominion, that presided over and structured the entire universe, Montesquieu, in *De l'esprit des lois*, gave law and regularity to the sublunary, historical realm – the realm of not only natural but also human activity, both producers of irregularity, imperfection and change. Like Newton, he discerned the laws of a particular dynamic system, not that of the motion of the planets, but that of the movement of political history. He made history into a science, an ordered body of knowledge structured by rules and governed by law.

However, for Montesquieu, history was not a science with laws simply because it had an identifiable order that evoked the ultimate law-giving dominion of God. In *De l'esprit des lois*, he did not explicitly name an all-powerful legislator and judge as the source of all laws, as Newton did in the *General scholium*. Rather, Montesquieu wrote in the first chapter that '[l]es lois, dans la signification la plus étendue, sont les rapports nécessaires qui dérivent de la nature des choses' ('[l]aws, taken in the broadest meaning, are the necessary relations deriving from the nature of things').[48] Natural laws derived from the nature (i.e. the essential, defining characteristic) of the world, divine laws from the nature of God, and political, historical laws from the nature of humankind.[49] Human actions were determined by both natural and divine law: 'L'homme, comme être physique, est, ainsi que les autres corps, gouverné par des lois invariables' ('Man, as a physical being, is governed by invariable laws like other bodies'). But Montesquieu's 'man' was more than just a 'physical being'; he was also an 'intelligent being', meaning that he did not merely describe the regularities he observed, but endeavored to understand why they were so. He disobeyed, intervened, and acted: 'Comme être intelligent, il viole sans cesse les lois que Dieu a établies, et change celles qu'il établit lui-même' ('As an intelligent being, he constantly violates the laws God has established and changes those he himself establishes').[50] In *De l'esprit des lois* people both followed the laws of nature and modified them to produce new (and often more useful, functional and better suited) laws, that is, new interfaces of interaction with their natural, social and political environments.

Montesquieu's work might at first appear deterministic. He famously identified three major forms of government and their natures, from which their laws flowed: democracy or rule by part or all of a people, monarchy or rule by a single person, and despotism or rule by a single person, who did not follow any laws, only his desires and whims.[51] Each of these governments functioned with respect to two larger principles, size and climate, which made them observable, comparable and explainable in human language. When speaking of despotic government, for example, Montesquieu wrote: 'Plus l'empire est étendu, plus le serail s'agrandit, et plus, par conséquent, le prince

48. Charles de Secondat, baron de La Brède et de Montesquieu, *De l'esprit des lois*, in *Œuvres complètes*, ed. Roger Caillois, 2 vols (Paris, 1949-1951) vol.2, p.227-995 (232). English translations of this work are from Montesquieu, *The Spirit of the laws*, ed. and translated by Anne M. Cohler, Basia Carolyn Miller and Harold Samuel Stone (Cambridge, 1989), here p.3.

49. Montesquieu writes that 'tous les êtres ont leurs lois; la Divinité a ses lois; le monde matériel a ses lois; [...] les bêtes ont leurs lois; l'homme a ses lois' ('all beings have their laws: the divinity has its laws, the material world has its laws, [...] the beasts have their laws, man has his laws'). Montesquieu, *De l'esprit des lois*, p.232; Montesquieu, *The Spirit of the laws*, p.3.

50. Montesquieu, *De l'esprit des lois*, p.234; Montesquieu, *The Spirit of the laws*, p.5.

51. Montesquieu, *De l'esprit des lois*, p.239.

est enivré de plaisirs. Ainsi, dans ces Etats, plus le prince a de peuples à gouverner, moins il pense au gouvernement' ('The more extensive the empire, the larger the seraglio, and the more, consequently, the prince is drunk with pleasures. Thus, in these states, the more people the prince has to govern, the less he thinks about government').[52] Climate also affected the nature of a government: differences in temperature changed 'le caractère de l'esprit et les passions du cœur' ('the character of the spirit and the passions of the heart'). A person in a cold place, Montesquieu thought, had 'plus de vigueur [...], plus de confiance en soi-même, c'est-à-dire plus de courage; plus de connoissance de sa supériorité, c'est-à dire moins de désir de la vengeance' ('men are more vigorous in cold climates [...]. [One has] more confidence in oneself, that is, more courage; better knowledge of one's superiority, that is, less desire for vengeance'). In contrast, people from a warm climate had 'peu de vivacité' ('little vivacity') but more 'sensibilité pour les plaisirs' ('sensitivity to pleasures').[53] (According to Montesquieu, this explained why despotic governments were to be found in hot places.)

Discovering the natural laws of politics not only explained how and why they functioned, but also enabled Montesquieu to train and develop readers' perceptive skills, as he did not consider these readers to be simply the passive subjects of natural law. Nature did not have to remain in its original, determined state; it could be improved. The fact that people could perceive law suggested that they could make law, signifying a relationship to the government which was not predetermined or already present in the natural course of its laws. Peoples' thoughts and desires could intervene in and change the physical, passionate and intellectual composition of the body politic. As Jean Goulemot argues in an analysis of Montesquieu's *Lettres persanes* (1721), the general laws of history – with respect to the size and climate of a country – and the natural laws of each type of government were always in dialogue, and sometimes even in conflict, with the laws invented by people over the course of history.[54] In each specific government the law was somewhat different, a product of peoples' interactions with and modifications of the natural system.

52. Montesquieu, *De l'esprit des lois*, p.250; Montesquieu, *The Spirit of the laws*, p.20.
53. Montesquieu, *De l'esprit des lois*, p.474-76; Montesquieu, *The Spirit of the laws*, p.231-34.
54. Jean Marie Goulemot, *Le Règne de l'histoire: discours historique et révolutions, XVIIe-XVIIIe siècle* (Paris, 1996), p.301-302.

The most productive connections between humans and law, according to Montesquieu, were those where people realized not only that they were active legislators, but that they could work to produce a more functional and more harmonious existence than before. *De l'esprit des lois* was filled with examples taken from history of attempts either successful (because they better regulated the passions of individuals and governments) or failed (because they pushed the members of the body of the state toward fear and decadence). Successful laws were the result of legislators who, according to Montesquieu, were able to perceive the natural economy of their government on a general level, and thus able to sense how its laws were hurting rather than benefiting the people. For instance, in chapter 6 of book 18, Montesquieu spoke of '[l]es anciens empereurs de la Chine' ('the former emperors of China') who, instead of forcing their people to live in – what Montesquieu deemed to be – a naturally hostile climate, employed 'l'industrie des hommes' ('the industry of men') to reshape the terrain, changing the relationship of the inhabitants to their climate, and thus the expression of the laws.[55] Not only did these emperors build upon a naturally watery and unwelcoming environment, constructing 'de dessous les eaux les deux plus belles provinces de l'empire' ('[t]he finest provinces in the empire were seen to rise from under the water'), but since their environmental modifications needed to be constantly maintained by the people, the government could no longer function through the laws of despotism, usually applied to large countries with lazy and selfish inhabitants. The human modification of the environment in turn required a dialogue with and a modification of natural law: 'Ainsi, malgré le climat de la Chine, où l'on est naturellement porté à l'obéissance servile, malgré les horreurs qui suivent la trop grande étendue d'un empire, les premiers législateurs de la Chine furent obligés de faire de très bonnes lois, et le gouvernement fut souvent obligé de les suivre.'[56] Here was an example – albeit, according to Montesquieu, unfortunately quite rare – of how being able to recognize the activity of natural law on the larger scale (with respect to the size of the state and its climate) 'obliged' people to move

55. Montesquieu, *De l'esprit des lois*, p.534; Montesquieu, *The Spirit of the laws*, p.288.
56. 'Thus, in spite of the climate of China, where one is by nature inclined to servile obedience, in spite of the horrors that attend an excessively large empire, the first legislators of China were obliged to make very good laws, and the government was often obliged to observe them.' Montesquieu, *De l'esprit des lois*, p.535; Montesquieu, *The Spirit of the laws*, p.288.

from a passive to an active state, intervening in and transforming what appeared at first to be purely deterministic conditions.

History for Montesquieu was not simply determined by a higher power. It was generated, ordered and guided by the interaction of human thought and activity with the laws natural to each form of government. There was no fixed point to which observed effects could be abstracted, no universal explanation for all events. The study of law, and thereby of history, needed to be approached from the point of view of its utility for people, who had to work to discover the underlying laws of their specific government and environment in order to understand what this utility meant. Law was no longer just an attribute of the natural world; Montesquieu defined it as historical, the product of interaction with human activity, entwined with the hope that eventually people's perspectives could become broad and general enough not only to see how the laws functioned but to grasp how they could make them function more smoothly, or perhaps change them altogether.

Scaling the levels of perception: the evolving relationship with nature in the *Histoire naturelle*

It is well known that Buffon admired Montesquieu's ideas and writing, and the two thinkers would probably have met in similar social and intellectual circles. The first volume of the *Histoire naturelle* was published only a year after *De l'esprit des lois*, and both authors had the experience of being censored by the Sorbonne. However, it would be mistaken to see Buffon as an imitator of Montesquieu; rather, as Jacques Roger has pointed out, he pushed Montesquieu's ideas on climate and natural law further.[57] Most importantly, Buffon drew upon the theory of political history as an interactive process in his study of natural history. The idea of an open-ended activity of modifying perception was what he borrowed from Montesquieu, formulating his own model of the discursive relationship between humans and their natural environments.

Over the course of the *Histoire naturelle*, Buffon presented more general and lyrical aspects of nature, expanding from the observation of the immediate moment and the surrounding world to the depths of

57. See Roger, *Buffon*, p.242-43. See also the article 'Buffon', in *Dictionnaire Montesquieu*, ed. Catherine Volpilhac-Auger (Lyon, 2013), http://dictionnaire-montesquieu.ens-lyon.fr/fr/article/1376426324/en (last accessed March 19, 2018), n.p.

time and space. Proceeding from the particular or local to the general was designed not only to demonstrate the workings of nature, but also to reveal Buffon's ability to discover these workings and to portray them in a language proportional to their generality and grandeur. In 'De la nature, première vue' ('On nature, first view'), the introductory essay to the twelfth volume (1764), he described the world as it could never be physically observed with the naked eye, aligning the majesty of the natural universe with the inventive power of human thought. In this text, he appeared to climb nearly to a divine point of view, and from here he illustrated the work and movement of nature as a whole. Controlled by the power of God, nature was described as the visible manifestation of the divine plan, a book whose pages were time and space, 'l'édifice du monde' ('the edifice of the world'), on which were inscribed the physical evidences of law. Nature molded space into 'des millions de globes opaques, [qui] circulans autour des premiers, [...] composent l'ordre et l'architecture mouvante [de l'univers]' ('millions of opaque globes [which], orbiting the original globes, [...] make up the order and moving architecture [of the universe]').[58] Nature was both God's work of art and the immortal worker: 'un ouvrage perpétuellement vivant, un ouvrier sans cesse actif [...]: le temps, l'espace et la matière sont ses moyens, l'Univers son objet, le mouvement et la vie son but' ('a perpetually living work, and a ceaselessly active worker [...]: time, space, and matter are its means, the Universe is its object, and movement and life are its goal').[59]

Natural history for Buffon was not simply a descriptive presentation of observations and beliefs about the natural world. Natural law was not merely a set of phenomena to be observed and affirmed. Producing natural history was a shared venture, combining the authority of nature and the creative ability of the author. In 'De la nature, première vue', Buffon seemed to peer over nature's shoulder, observing the writing of its story. His language, in turn, transmitted and rendered legible to readers the invisible mechanics or hidden power of the world: not only its order, but the procedure and process of its creation. One year later, in 'De la nature, seconde vue', where his subject was not the motion of the planets but the definition of species of animals, Buffon continued to model an ascent of the ladder of both existence and perception. He wrote not of a group or chain of individuals composing a species (as he had in the 1750s), but of a

58. Buffon, *Histoire naturelle* (1749-1767), vol.12, p.v-vi.
59. Buffon, *Histoire naturelle* (1749-1767), vol.12, p.iii.

species itself as an individual, presenting a more general perspective which communicated more efficiently and powerfully the order of nature:

> Un individu, de quelque espèce qu'il soit, n'est rien dans l'Univers; cent individus, mille ne sont encore rien: les espèces sont les seuls êtres de la Nature; êtres perpétuels, aussi anciens, aussi permanens qu'elle; que pour mieux juger nous ne considérons plus comme une collection ou une suite d'individus semblables; mais comme un tout indépendant du nombre, indépendant du temps; un tout toujours vivant, toujours le même.
>
> An individual, from any species, is nothing in the Universe; one hundred individuals, even one thousand are still nothing: species are the only [real] beings in nature; perpetual beings, as ancient and as permanent as she, beings that, to better judge, we consider only as a collection or a series of similar individuals. But they are a whole independent of number, independent of time; a whole which is always living and always the same.[60]

Once seen as disorderly and chaotic, as fragments of an indiscernible whole (images from the first volume of the *Histoire naturelle*, which will be discussed in Chapter 3), nature appeared to change at this new level of insight and with this new ability of language. The observer portrayed in this passage was now an individual in a new sense: 'l'espèce humaine entière' ('the entire human species'), the witness to all of nature's story through the ages.[61] Buffon asked his readers to reflect upon the world they had learned to read many times over in his text, and then to take the next logical step in perception, selecting and generalizing from their observations. A single person, he wrote, saw nature die in winter and become reborn in spring. Man, the species, was capable of combining the experiences of hundreds of seasons over the centuries:

> Mais ces idées de renouvellement et de destruction, ou plutôt ces images de la mort et de la vie, quelque grandes, quelque générales qu'elles nous paroissent, ne sont qu'individuelles et particulières; l'homme, comme individu, juge ainsi la Nature, l'être que nous avons mis à la place de l'espèce la juge plus grandement, plus généralement; il ne voit dans cette destruction, dans ce renouvellement, dans toutes ces

60. Buffon, *Histoire naturelle* (1749-1767), vol.13, p.i.
61. Buffon, *Histoire naturelle* (1749-1767), vol.13, p.iv.

> successions que permanence et durée; la saison d'une année est pour lui la même que celle de l'année précédente, la même que celle de tous les siècles; le millième animal dans l'ordre des générations, est pour lui le même que le premier animal. Et en effet, si nous vivons, si nous subsistons à jamais, si tous les êtres qui nous environnent subsistoient aussi tels qu'ils sont pour toujours, et que tout fût perpétuellement comme tout est aujourd'hui, l'idée du temps s'évanouiroit et l'individu deviendroit l'espèce. Eh pourquoi nous refuserions-nous de considérer la Nature pendant quelques instans sous ce nouvel aspect?

> Yet these ideas of renewal and destruction, or rather these images of death and of life, as grand and as general as they seem to us, are but individual and particular: man, as an individual, thus judges Nature, the being that we have put in the place of species judges it more grandly and generally. He does not see, in this destruction, in this renewal, in all these successions, but permanence and duration; the season of one year is for him the same as that of the preceding year, the same as that of all the centuries; the thousandth animal in the order of generations is for him the same as the first animal. And indeed, if we live, if we subsist forever, if all the beings that surround us also subsisted just as they are forever, and everything were perpetually as everything is today, the idea of time would vanish and the individual would become the species. So why would we refuse to consider Nature, for a little while, from this new angle?[62]

The above passage exemplifies how Buffon thought that natural history should strive to articulate the order and energy of the world. Knowledge of natural history at this level could no longer be obtained through direct observation. The reader became, at this stage, more dependent upon the lyrical quality and spirit of Buffon's discourse in order to truly understand the world.

Buffon's work drew upon Montesquieu's in the sense that his lyricism was not only meant to transmit a broader and more complete comprehension of nature. With an increasingly general perception of the world the reader moreover gained the ability to invent, innovate and change the workings of nature in order to render it more beautiful and more useful to people. Creativity was communicated through the ability to perceive on a larger scale both the laws of nature and the history of human interaction with nature. When appreciated from the perspective of the entire species of mankind, history resolved into the stages of perception that allowed for the current one. These levels were

62. Buffon, *Histoire naturelle* (1749-1767), vol.13, p.ii-iii.

the intellectual markers of the development of human thought and its gradual interaction with the natural world:

> [L]'homme en venant au monde arrive des ténèbres; l'ame aussi nue que le corps, il naît sans connoissance comme sans défence, il n'apporte que des qualités passives, il ne peut que recevoir les impressions des objects et laisser affecter ses organes, la lumière brille long-temps à ses yeux avant que de l'éclairer: d'abord il reçoit tout de la Nature et ne lui rend rien, mais dès que ses sens sont affermis, dès qu'il peut comparer ses sensations, il se réfléchit vers l'Univers, il forme des idées, il les conserve, les étend, les combine; l'homme, et sur-tout l'homme instruit, n'est plus un simple individu, il représente en grande partie l'espèce humaine entière.

> [M]an, coming into the world, arrives from the abyss; the soul as naked as the body, he is born without knowledge as without defense, he only brings passive qualities, he can only receive impressions of objects and let his organs be affected, light shines for a long time in his eyes before enlightening him. At first, he receives everything from Nature and gives back nothing, but as soon as his senses are strengthened, as soon as he can compare sensations, he reflects toward the Universe, he forms ideas, conserves them, stretches them, combines them; man, and especially the educated man, is no longer a simple individual, he represents to a large extent the entire human species.[63]

From a shadowy, helpless beginning, Buffon imagined how people became passive observers, and then developed active minds capable of combining the order of their reason with the order of nature, bringing their thought into harmony with the spirit and operation of natural laws. Buffon's narrative in the *Histoire naturelle* not only showed his readers the history of their interaction with the world, but also hinted at their capacity to continue the dialogue with nature, deepening and widening the scope of their perception.

Buffon intended to use language as a tool of measurement, mirroring the laws of nature in the rules of perception, allowing both the phenomenal and intellectual worlds to be perceived from above. Having given his readers this privileged perspective, Buffon then asked them to use the spirit or energy transmitted to them through reading his language to make nature better, more functional and more beautiful. Buffon explained to readers their role in the design of the

63. Buffon, *Histoire naturelle* (1749-1767), vol.13, p.iii-iv.

universe: man was here 'pour seconder la Nature [...]; seul entre tous, capable de connoître et digne d'admirer, Dieu l'a fait spectateur de l'Univers et témoin de ses merveilles' ('to assist Nature [...]; the only one capable of knowing and admiring it, God made him a spectator of the Universe and a witness to its marvels').[64] Divine light, filtered through Buffon's text, permitted man to study the world; through this light, 'il voit et lit dans le livre du monde' ('he sees and reads from the book of the world'). As he studied nature, guided by the energy of Buffon's language, his ability to understand it increased, and he 's'élève par degrés au trône intérieur de la toute-puissance' ('rises by degrees to the inner throne of [God's] omnipotence').[65]

In Buffon's story of nature, people did more than observe, contemplate and know. They were also the only beings who could modify nature and thus who could reform the course of history. For example, in their interaction with other animals, especially those in closest proximity and of greatest practical use (cows, horses, donkeys, dogs, cats), they changed species, defining them by how they fit human needs, transforming them from innovative ideas in the mind to realities in the world: 'nous avons altéré, modifié, changé [les espèces] des animaux domestiques: nous avons donc fait des genres physiques et réels [...]; ces genres physiques sont réellement composés de toutes les espèces que nous avons maniées, modifiées et changées' ('we have altered, modified, changed [species] of domesticated animals: we have thus created physical and real categories [...]; these physical categories are actually made up of all the species we have manipulated, modified and changed').[66] These species no longer existed as nature had created them, yet they remained nonetheless part of the natural order; human changes were not unnatural nor should they be viewed as outside of nature's domain. To 'assist' ('seconder') nature meant to continue the process of its production – a process of which people could become aware by learning to observe its laws, through the reading of the *Histoire naturelle.*

When it could be perceived from above, Buffon believed that nature became beautiful, and this beauty was in no small part due to how people used their understanding of natural law to actively change the structure and operation of the world. Buffon's images of nature in its original state, untouched by people, were quite

64. Buffon, *Histoire naturelle* (1749-1767), vol.12, p.xi.
65. Buffon, *Histoire naturelle* (1749-1767), vol.12, p.xi.
66. Buffon, *Histoire naturelle* (1749-1767), vol.11, p.369.

negative, even projecting a moralizing tone that highlighted nature's impurity in this form. He passionately described a planet 'surmontée par les débris de ses productions, [qui] n'offre au lieu d'une verdure florissante, qu'un espace encombré, traversé de vieux arbres chargés de plantes parasites, de lichens, d'agarics, fruits impurs de la corruption' ('conquered by the debris of its productions [which], instead of offering a flourishing greenery, is but a congested space, strewn with old trees laden with parasitic plants, lichens, fungi, the impure fruits of corruption'). There were only 'des eaux mortes et croupissantes' ('dead and stagnant waters'), 'des terrains fangeux' ('miry ground'), '[d]es forêts décrépites' ('decrepit forests') and 'des marécages qui couverts de plantes aquatiques et fétides, ne nourrissent que des insectes vénéneux et servent de repaire aux animaux immondes' ('swamps which, covered with aquatic and fetid plants, only nourish poisonous insects and serve as lairs for vile animals').[67] Yet this stagnant, rotting and filthy nature could be saved: after the order of their ideas had been changed by Buffon's text, people could use this new perceptive ability to render nature better than it was, creating a beauty and a utility that had not been there before:

> Nulle route, nulle communication, nul vestige d'intelligence dans ces lieux sauvages; l'homme obligé de suivre les sentiers de la bête farouche, s'il veut les parcourir; contraint de veiller sans cesse pour éviter d'en devenir la proie; effrayé de leurs rugissemens, saisi du silence même de ces profondes solitudes, il rebrousse chemin et dit: la Nature brute est hideuse et mourante; c'est Moi, Moi seul qui peux la rendre agréable et vivante: desséchons ces marais, animons ces eaux mortes en les faisant couler, formons-en des ruisseaux, des canaux; employons cet élément actif et dévorant qu'on nous avoit caché et que nous ne devons qu'à nous-mêmes; […] une Nature nouvelle va sortir de nos mains.

> No roads, no communication, no vestige of intelligence in these savage places; if he wants to travel them, man [is] obliged to follow the trails of the savage beast; forced to keep endless vigil to avoid becoming their prey; scared by their roaring, even struck by the silence of these profound solitudes, he retraces his steps and says, 'brute Nature is hideous and dying; it is I, I alone who can render it agreeable and lively: let us dry up these swamps, animate these dead waters by making them flow, let us make streams, canals; let us use that active and all-consuming element that was hidden from us and

67. Buffon, *Histoire naturelle* (1749-1767), vol.12, p.xii.

> that we can only attribute to ourselves; [...] a new Nature will spring from our hands.[68]

Buffon began this passage with 'l'homme', a solitary being overpowered by nature, and then transitioned to a singular emphatic pronoun, 'Moi', the conscience of mankind as it started to rework nature. But this singular conscience was not complete until it could blend with the authorial and oratorical voice of Buffon, written in the first-person plural of the verb: 'desséchons', 'animons', 'employons'. Through Buffon's text people received creative energy, heat, 'cet élément actif' which not only transmitted understanding but moderated the world, changing its terrifying, ugly, chaotic parts into comprehensible and measurable entities, allowing the formation of a new, more beautiful world from 'our hands', because of human spirit, and hence from the rational capacity of the human mind. It is this idea of the heat of expression, the production of a dynamic language that mirrored both the experience and ideas of the mind and the levels of nature, that will be addressed in the following chapter.

68. Buffon, *Histoire naturelle* (1749-1767), vol.12, p.xii-xiii.

2. Generating heat: the energy of natural language

'Je le ferai brûler dans un creuset d'or', disait Guyton de Morveau, pour s'assurer que le diamant contenait de la terre fixe. 'Le meilleur creuset, c'est l'esprit', répondit Buffon.

'I will burn it in a golden crucible', said Guyton de Morveau, in order to be certain that diamonds contain oxides. 'The best crucible is the mind', replied Buffon.[1]

[T]ant d'ouvrages qui ne seront appuyés que sur un froid bon sens & sur une pesante raison, seront peut-être fort estimés, mais peu lûs, & tomberont enfin dans l'oubli, lorsqu'un homme doüé d'un beau génie & d'une grande éloquence les aura dépouillés, & qu'il aura reproduit aux yeux des hommes des vérités, auparavant d'une austérité seche & rebutante, sous un vêtement plus noble, plus élégant, plus riche & plus séduisant.

[S]o many works that will be based only upon a cold common sense and a weighty reason, will perhaps be very esteemed, but infrequently read, and will finally fall into oblivion, when a man gifted with a great genius and eloquence will have scrutinized them, and will have reproduced for the eyes of all truths, formerly

1. Quoted in Daniel Mornet, *Les Sciences de la nature en France au XVIIIe siècle* (Geneva, 2001), p.114.

> of a dry and repulsive austerity, in far more
> noble, elegant, rich and seductive robes.[2]

How did the rational, mathematical foundations of thought, perception and observation of nature become the basis for the reproduction of nature through Buffon's language? The *Histoire naturelle* is an example of how this Enlightenment scientific activity must be understood through its discursive, written form. Natural history was the human reading and interpretation of a well-measured world, all parts of which were proportional to all others, not just at the level of immediate surroundings and present moment, but on every level, every scale of perception, at every moment of measurable time. The science of natural history was the ordering and translation of the proportions and relationships of nature into words. For Buffon, written language was the only way to truly articulate the world's operation, the generality and ubiquity of its laws. Thus, what made his work 'natural' was not only the communication and harmonization of the orders of thought and nature, but furthermore his belief that to embody the science of natural history the words of nature could be none other than eloquent, powerful, imbued with the ability of nature itself.

Introducing heat: *De l'art d'écrire*

In *De l'art d'écrire* (*On the art of writing*), Buffon described the process of translating the natural world into language and transmitting this new, literary world to readers.[3] He argued that nature was best depicted through what he called 'painting', which not only represented its order, but made its parts flow together in such a way that they resembled and transmitted life itself:

> Peindre ou décrire sont deux choses différentes: l'une ne suppose que des yeux, l'autre exige du génie [...]. La description présente

2. Denis Diderot, 'Encyclopédie', in Diderot and D'Alembert, *Encyclopédie*, vol.5, p.638.
3. This text was probably part of the generation of the 'Discours sur le style' (1753). Although not mentioned in Jacques Roger and E. Genet Varcins' 'Bibliographie de Buffon', this piece was published in Buffon's correspondence. It was copied from Buffon's notes and kept by his notary. Henri Nadault de Buffon, *Correspondance inédite de Buffon*, vol.1 (Paris, 1860), p.292-94. The modern edition of this text that I use here is Georges-Louis Leclerc de Buffon, *De l'art d'écrire* (Castelnau-le-Lez, 1992), p.37-41.

successivement et froidement toutes les parties de l'objet; plus elle est détaillée, moins elle fait d'effet. La peinture au contraire, ne saissisant d'abord que les traits les plus saillants, garde l'empreinte de l'objet et lui donne de la vie. Pour bien décrire, il suffit de voir froidement; mais pour peindre, il faut l'emploi de tous les sens. Voir, entendre, palper, sentir, ce sont autant de caractères que l'écrivain doit sentir et rendre par des traits énergiques. Il doit joindre la finesse des couleurs à la vigueur du pinceau, les nuancer, les condenser ou les fondre; former enfin un ensemble vivant, dont la description ne peut présenter que des parties mortes et détachées.

Painting and description are two very different things. The latter considers only the eyes, the former demands genius [...]. Description coldly and sequentially shows all the parts of an object; the more detailed it is, the less effect it has. Painting, by contrast, seizes first upon the most salient traits, retaining the imprint of the object and giving it life. In order to write a good description, cold observation suffices; but painting requires the use of all the senses. Seeing, hearing, touching, feeling: these are all elements which the writer must sense and render in energetic strokes. He must link the finesse of color and the vigor of the paintbrush, nuancing, condensing or melting these colors, and ultimately shaping a living ensemble of which description could only have presented dead and detached parts.[4]

In contrast to the activity and liveliness of painted language, a purely descriptive work was like a sketch that lacked shading, depth and thus vitality – as indicated by the recurrence of the expressions 'cold' and 'coldly'. Description sapped the vivacity of nature, producing a lifeless text of 'dead and detached parts'.

Buffon's contrast between 'description' and 'painting' when characterizing his writing about nature was not only borrowed from the visual arts. It also referred to a convention in historical writing (discussed in the introduction) at the time which stipulated that history only describe and enumerate events, without attempting to extrapolate more general meaning and relationships between them.[5] In rethinking the discipline of natural history, Buffon sought to combine history

4. Buffon, *De l'art d'écrire*, p.38-39.

5. According to Elizabeth Liebman, an important influence on Buffon's ideas in *De l'art d'écrire* was André Féliben's 'classification of the elements of painting'. Here Féliben 'prioritized composition and considered drawing and coloring to be only mechanical practice. The painter's achievement in the "grand manner" was the generation and ordering of mental images' (Liebman, 'Painting natures', p.18).

with physics, or the search for causes, moving from specific, detailed observations of events and phenomena to a narrative of the large-scale activities of nature as it moved through vast blocks of time. The term 'painting' thus allowed him to articulate what he wanted his language to do: reproduce the temporal energy of nature on the page. He argued that a text which employed 'painting' was warmer, richer and more animated, exuding this energy of nature as it worked through time. In fact, he seemed to think that words could render ideas even more visible than could an actual painting: a talented writer's 'phrases et ses mots feront plus d'effet que le pinceau et les couleurs du peintre' ('sentences and words will have more effect than the paintbrush and the colors of the artist').[6] The painter of words could not only see, but also 'entendre, palper, sentir' ('hear, touch, feel'). He wrote with both passion and intellect. His mind 'saisit' ('[seizes]') upon nature; Buffon used verbs such as 'joindre', 'nuancer', 'condenser', 'fondre' ('to melt'), 'former' ('to shape') and 'rendre' to illustrate the activity of combining parts into a new whole, the text, which seemed to be just as alive as the world it purported to reenact.

Buffon's language also revealed the ability to perceive an object from many different angles: 'l'emploi de tous les sens' ('the use of all the senses') to portray the world with 'des traits énergiques' ('energetic strokes'). The activity of painting captured not only the exterior form of an object, but the communication between all its parts, the process through which they worked together and the relationship of this individual body to the surrounding world (past and present), culminating in 'un ensemble vivant' ('a living ensemble') and an increasingly general perspective of nature.[7] Buffon characterized the transmission of nature's energetic movement through the medium of language as having heat: 'Le grand écrivain peut [...] *échauffer, embraser* son lecteur par la représentation de plusieurs actions qui toutes auront de la *chaleur*, et qui par leur union et leurs *rayons se graveront* dans sa mémoire et subsisteront indépendamment de l'objet.'[8] Reading a

6. Buffon, *De l'art d'écrire*, p.39. On the tradition of 'painting with words' and the relationship between rhetorical and visual composition, see Angelica Goodden, *Actio and persuasion: dramatic performance in eighteenth-century France* (Oxford, 1986), p.11-12.
7. Buffon, *De l'art d'écrire*, p.39-40.
8. 'The great writer can [...] heat up, set alight, his reader through the representation of several activities that all have heat, and that, by their union and their rays, engrave themselves in the reader's memory and live on independently of the [original] object.' Buffon, *De l'art d'écrire*, p.40-41 (my emphasis).

nature that was not merely described but indeed painted was meant to set the reader's mind ablaze in a manner more powerful than direct observation or cold, lifeless description. Buffon wanted to leave readers with a lasting 'heat', which would perpetuate and 'engrave' in their memories nature's phenomena, order and laws – its history in a new sense. The march of natural history and the experience of its observation were restaged and harmonized on the page.

Heat: the material interface with nature

When Buffon composed *De l'art d'écrire* in the 1750s the heat produced by painting with words acted as a metaphor to conceptualize the function of prose as a tool for the reproduction of natural knowledge on the page. However, even earlier in his career, he had taken an interest in the material heat of nature, in the study of the mechanisms of life and vitality, and what made some things alive while others were not. Such a question was deeply rooted in chemical, mechanical and medical philosophies of the time, and an examination of these theories will show how discursive heat could be understood as a form of physical heat, and how the combination of these two types of heat contributed to the definition and practice of Buffon's discipline of natural history.

Buffon's 1735 publication of *La Statique des végétaux et l'analyse de l'air*, a translation of Stephen Hales' *Vegetable staticks* (1727) allowed him to speculate upon the movement and circulation of nutrients and other organic compounds in living bodies. Hales' work was a foundational essay in the area of plant physiology as well as an example of the application of Newtonian physics to the study of living nature. It drew on the work of physiologists of the seventeenth century such as William Harvey, who had studied the circulation of blood in animals, in order to understand the movement of sap within plants as well as their material exchange with the surrounding environment.[9] Buffon's interest in Hales' experiments on plants is reflected in a larger philosophical approach developed over the course of the *Histoire naturelle* in which he, too, applied Newton's mechanics of the

9. Richard S. Westfall, 'Hales, Stephen', in *Catalogue of the scientific community of the 16th and 17th centuries*, http://users.clas.ufl.edu/ufhatch/pages/03-Sci-Rev/SCI-REV-Home/resource-ref-read/major-minor-ind/westfall-dsb/index.htm (last accessed March 19, 2018), n.p. On Hales also see Thomas Hankins, *Science and the Enlightenment* (Cambridge, 1985), p.50-53.

attraction and repulsion of particles (famously discussed in the thirty-first 'Query' of the *Opticks*) to the activity and movement of living things.[10] In his translation of Hales the fundamental operation of life was interpreted as an attractive force that held bodies together, giving them form and most of their identifiable, external qualities:

> Nous trouvons par l'analyse chymique des Végétaux, qu'ils sont composés de soulfre, de sels volatils, d'eau, de terre & d'air; ces quatre premiers principes agissent les uns sur les autres par une forte puissance d'attraction mutuelle, & l'air que je regarde comme le cinquième principe, est doué de cette même puissance d'attraction, lorsqu'il est dans un état fixe; mais il exerce la puissance contraire aussi-tôt qu'il change d'état; car dèslors il repousse avec une force supérieure à toutes les forces connues. Tout se fait donc dans la nature par la combinaison de ces cinq principes, par leur action & réaction réciproque.
>
> We find by the chymical analysis of vegetables, that their substance is composed of sulpher, volatile salt, water and earth; which principles are all endued with mutually attracting powers, and also of a large portion of air, which has a wonderful property of strongly attracting in a fixt state, or of repelling in an elastick state, with a power which is superior to vast compressing forces; and it is by the infinite combinations, action and re-action of these principles, that all the opertions in animal and vegetable bodies are effected.[11]

Yet this attractive force could not alone explain the essence of life: an opposing force, in the above citation 'air', was understood to resist the pull of attraction, generating the necessary space and pressure for sap, blood and water to move through the body, allowing for a vital economy that nourished and promoted growth. As Violeta Aréchiga has shown, for Hales the idea of air was associated with life; once air

10. On the application of mechanical philosophy to the study of life see Theodore M. Brown, 'From mechanism to vitalism in eighteenth-century English physiology', *Journal of the history of biology* 7:2 (autumn 1974), p.179-216 (185-89); Thomas S. Hall, 'On biological analogs of Newtonian paradigms', *Philosophy of science* 35 (1968), p.6-27; Jacques Roger, 'Chimie et biologie: des "molécules organiques" de Buffon à la "physico-chimie" de Lamarck', *History and philosophy of the life sciences* 1 (1979), p.43-64; and Buffon, *Histoire naturelle*, ed. Schmitt and Crémière, 'Introduction', vol.2 (2008), p.7-90 (23-25).
11. Stephen Hales, *La Statique des végétaux et l'analyse de l'air* […] *traduit de l'anglais par M. de Buffon, de l'Académie royale des sciences* (Paris, Debure l'aîné, 1735), p.270. Hales, *Vegetable staticks* (London, W. and J. Innys, 1727), p.318-19.

entered into a solid body, it lost its 'elasticity' or expansive power, and 'se retrouvait fixé dans les substances animals, végétales et minérales. En fait, à l'état fixe, l'air formait la matière, ductile et nutritive, à partir de laquelle les organes végétaux croissaient' ('found itself fixed in animal, vegetal and mineral substances. Indeed, in this fixed state, air made up the ductile and nutritive matter from which the organs of the plant grew').[12] Buffon translated Hales' quotation of Newton, which described the 'principle' of air as '*une nourriture tendre & humide [qui] est aisément disposée par une chaleur douce, & un mouvement temperé, à changer de forme & de contexture; les mouvemens intestins rassemblant les particules homogènes, & séparant les particules étherogènes.* Newton Optique, qu. 31' ('This soft and moist nourishment easily changing its texture by gentle heat and motion, which congregates homogeneal bodies, and separates heterogeneal ones. *Newton's Opticks*, qu. 31').[13] The life of a being was characterized by the interaction of 'homogeneal' particles, gravitating to the center of the organism, with 'heterogeneal' ones (which Buffon appeared, interestingly, to have translated as 'étherogènes' and not 'héterogènes', as Hales originally wrote it), tending toward the outside, the ether. These latter particles of captured air were also called 'fixed fire' and provided the heat necessary for movement and change in living beings.

This example of heat in living beings may be viewed in the larger context of Enlightenment debates about the meaning and mechanics of fire and heat. In his *Encyclopédie* article on fire, D'Alembert began with a definition that seemed to raise more questions than it answered:

> Le caractere le plus essentiel du *feu*, celui que tout le monde lui reconnoît, est de donner de la chaleur. Ainsi on peut définir en général le *feu*, la matiere qui par son action produit immédiatement la chaleur en nous. Mais le *feu* est-il une matiere particuliere? ou n'est-ce que la matiere des corps mise en mouvement? c'est sur quoi les Philosophes sont partagés.
>
> The most essential characteristic of fire, the one that everyone recognizes, is to give heat. Thus, fire may be defined generally as the matter, which by its action, immediately produces heat in us. But is

12. Violeta Aréchiga, 'Le feu et la vie: la pensée chimique de Buffon', in *L'Héritage de Buffon*, ed. Marie-Odile Bernez (Dijon, 2009), p.235-44 (237). On other early modern theories of animal heat, see Everett Mendelsohn, *Heat and life: the development of the theory of animal heat* (Cambridge, MA, 1964).
13. Hales, *La Statique des végétaux*, p.271. Hales, *Vegetable staticks*, p.319.

> fire a specific kind of matter? Or is it but the matter of bodies put into motion? On this subject, philosophers are divided.[14]

Fire was that which produced the physical sensation of heat to which people can attest. Yet this definition did not explain what fire was or how it operated. Was it part of bodies or was it a substance that moved through bodies, creating flames, heat, light?[15] The more fundamental question of how to define a body was thus raised. Were bodies, and the matter of which they were composed, defined through Aristotelian physical qualities: hot, dry, wet, cold? Were they defined through extension? Or were they explained by how they changed – and, if so, what made them change?

Heat was a way of speaking of change brought about by movement. On the one hand, heat could be understood as the product of a kind of fire, an 'element', inherent to the substance of matter and present to differing degrees in all bodies. On the other hand, fire was what gave form to matter or altered matter by moving through it. It was thus a tool of 'analysis' used to understand the composition of bodies.[16] In this second sense, fire either could be construed as a mechanical process – experienced and described as heat – that made the particles of a body move faster, thus changing the state of the matter; or it could be explained as a 'quintessence', 'a celestial fire that embeds itself in the heart of matter' creating life, light and energy in bodies.[17]

At the beginning of the eighteenth century in France, fire and heat could be interpreted mechanically, through what Bernard Joly calls 'un cartésianisme corrigé par Malebranche' ('a Cartesianism modified by Malebranche'), in which fire was 'une agitation de la matière' ('an

14. Jean D'Alembert, 'Feu', in Diderot and D'Alembert, *Encyclopédie*, vol.6, p.599. At the beginning of this article, readers are recommended to consult the articles 'Cartésianisme', 'Matière subtile' (the Newtonian explanation of fire as a subtle fluid) and 'Chaleur'.
15. On the different types and uses of fire – as opposed to phlogiston, the 'element of fire' – in Enlightenment chemistry, see Christine Lehman, 'Gabriel François Venel (1723-1775): sa place dans la chimie française du XVIIIe siècle', doctoral dissertation, Université de Paris-Nanterre, 2006, especially ch.4, in which Lehman analyzes the chemist Venel's treatise 'Du feu' ('On fire'). See also Hélène Metzger, 'La théorie du feu d'après Boerhaave', *Revue philosophique de la France et de l'étranger* 109 (January-June 1930), p.253-85.
16. Bernard Joly, 'Les théories du feu de Voltaire et de Mme Du Châtelet', *SVEC* 2001:11, p.212-37 (214).
17. '[U]n feu céleste venu loger au cœur de la matière'. Joly, 'Les théories du feu', p.214.

agitation of matter') that caused its vortex-like particles (*tourbillons*) to move faster and to hit against one another, creating sparks perceived as light and heat.[18] Life, too, could be explained by means of this Cartesian mechanical theory, also called iatromechanics: its heat was created and rendered sensible through, in the words of Jacques Roger, 'les structures particulières des machines vivantes' ('the unique structures of living machines').[19] The Newtonian theory of fire, particularly interested in the emission of light from bodies, added universal attraction to the mix, using this force to explain how particles of light entered into bodies and changed their state, rendering them hot and luminous.[20] The terms 'subtle fluid' or 'spirit' of heat were often employed in chemistry, and could also be applied to physical theories of heat to explain how a force of life kept bodies moving and growing, in opposition to the compacting, organizing tendency of gravitation.[21] Phillip Sloan explains how this dialogue between attraction and heat moved beyond the purely mechanical, allowing nature to be understood 'as a creative and dynamic agency', which wrote a history of both universality and change.[22]

The theorization of heat as a 'subtle fluid', or as part of an 'active ether', enabled natural philosophers of the 1740s such as Buffon to create their own combinations of mechanics and the laws of attraction or gravitation. Buffon applied these ideas to the study of

18. Bernard Joly, 'La question de la nature du feu dans la chimie de la première moitié du XVIIIe siècle', *Corpus* 36 (1999), p.41-63 (43-44, 49). Descartes's explanation of fire can be found in the third part of *Le Monde, ou Traité de la lumière* (published in 1667), in which he explains the development of bodies through the physical shape, size and movement of particles.
19. Roger, 'Chimie et biologie', p.44. On iatromechanics, see also Anne C. Vila, *Enlightenment and pathology: sensibility in the literature and medicine of eighteenth-century France* (Baltimore, MD, 1998), p.15-17.
20. Hélène Metzger, 'Newton: la théorie de l'émission de la lumière dans la doctrine chimique au XVIIIe siècle', *Archeion* 11 (1929), p.13-25 (17). For an updated and more nuanced study of the theorization of light, see Lawrence M. Principe, 'Wilhelm Homberg et la chimie de la lumière', *Methodos: savoirs et textes* 8 (2008), http://methodos.revues.org/1223 (last accessed March 19, 2018), n.p.
21. This latter interpretation of an almost divine spirit keeping matter from compacting under the force of attraction was held by the marquise Du Châtelet, in her 1742 *Dissertation sur la nature et la propagation du feu*. Voltaire, by contrast, decided to take a more mitigated position, claiming that fire held an intermediate status between matter and spirit. Joly, 'La question de la nature du feu', p.43, 53-55.
22. Sloan, 'Natural history', p.301-302.

both inorganic and organic nature, defining life not as distinct from death, but as the degree of fire or heat in a body. In 1749, he wrote in the 'Comparaison des animaux et des végétaux' that 'il n'y a aucune différence absolument essentielle et générale entre les animaux et les végétaux, mais que la nature descende par degrés et par nuances imperceptibles d'un animal qui nous paroît le plus parfait à celui qui l'est le moins, et de celui-ci au végétal' ('there is no absolutely essential and general difference between animals and plants, but [rather] that nature descends by degrees and imperceptible nuances from the animal that to us seems the most perfect [humans] to that which is the least perfect, and from there to the vegetal').[23] He held that degrees of life and heat continued into the realm of the mineral which contained, in the form of soil and fossils, for example, the traces of previously 'organized' beings and thus the potential to build new life forms.[24] It was therefore possible for fire and heat to occupy an intermediary status, defining the organization of matter and thereby its degree of life, and, flowing in between bodies – physical, celestial and indeed, as the next section will show, textual – changing them.[25] In Peter Reill's words, as a fluid, heat came to be imagined as 'liquid' and thus 'entailed the penetration of the solid by an active, subtle fluid that energized the solid, releasing it from its cold imprisonment'.[26] In this sense, heat became a mode of communication, a cause of interaction. It engendered a perception of bodies of matter not as how they differed from one another, but as related, receptive, continuous.

Buffon took the interrelatedness of bodies to its most general limit. In her study of Enlightenment chemistry, Hélène Metzger quotes Buffon's 'De la nature, seconde vue', and his explanation of applying Newtonian laws to the study of chemistry. In this essay, Buffon wrote that:

> Newton a bien soupçonné que les affinités chimiques, qui ne sont autre chose que les attractions particulières dont nous venons de parler, se faisoient par des loix assez semblables à celles de la gravitation; mais il ne paroît pas avoir vu que toutes ces loix particulières n'étoient que

23. Buffon, *Histoire naturelle* (1749-1767), vol.2, p.8-9.
24. Buffon, *Histoire naturelle des minéraux*, 5 vols (Paris, Imprimerie royale, 1783-1788) vol.1, p.4-5.
25. Hankins, *Science and the Enlightenment*, p.50-53 and 72-75. This theory was applied to the mechanical as well as to the chemical sciences. See Metzger, 'Newton: la théorie de l'émission de la lumière', p.17.
26. Reill, *Vitalizing nature in the Enlightenment*, p.79-80.

> de simples modifications de la loi générale, et qu'elles n'en paroissoient différentes que parce qu'à une très-petite distance la figure des atômes qui s'attirent, fait autant et plus que la masse pour l'expression de la loi, cette figure entrant alors pour beaucoup dans l'élément de la distance. C'est cependant à cette théorie que tient la connoissance intime de la composition des corps bruts.
>
> Newton rightly suspected that chemical affinities, which are nothing else but the specific attractions about which we just spoke, were made by laws quite similar to those of gravitation; but he did not seem to have seen that all of these specific laws were nothing but simple modifications of the general law, and that they only appeared different because at a very small distance the shape of attracting atoms changes the expression of the law as much as[,] if not more[,] than their mass, which figures significantly in the element of distance. It is, however, this theory upon which the knowledge of the composition of inorganic bodies is based.[27]

Metzger uses this passage from the *Histoire naturelle* to show the extent to which Buffon, as he developed his discipline of natural history, worked to subsume chemistry and mechanics under the generality of natural law, which he considered universal in both its extent and its expression (just as he did in his debate with Clairaut). The applicability of the law of attraction (and the corresponding idea of repulsion that Buffon added) across the study of nature – from the perfection of the heavens to the sublunary realm of change, and from the disposition and relationships of ideas in the mind to the structure and operation of the outside world – meant that the universe could be understood and represented through the same rules and hence the same language of mechanics and heat. Although she describes how Buffon's way of thinking was ultimately rejected in the study of chemistry, Metzger writes that 'la loi d'*attraction universelle* rattachait directement les lois d'affinités spéciales […] à la conception générale du monde dont l'unité enchantait le théoricien' ('the law of *universal attraction* directly linked laws of special affinities to the general conception of the world whose unity enchanted theorists') and, borrowing again Buffon's words, she describes how

27. Buffon, *Histoire naturelle* (1749-1767), vol.13, p.xiv; quoted in Hélène Metzger, *Newton, Stahl, Boerhaave et la doctrine chimique* (Paris, 1974), p.58-59. I was directed to Metzger's analysis of this passage from a footnote in Arthur Donovan, 'Buffon, Lavoisier, and the transformation of French chemistry', in *Buffon 88: actes du colloque international*, ed. Jean-Claude Beaune *et al.* (Paris, 1992), p.387-95 (390).

'la méthode newtonienne [...] "n'abandonnait jamais l'analogie de la nature"' ('the Newtonian method [...] "never abandoned the analogy of nature"').[28] Heat, flowing through individual bodies, put them into relationship with one another, and endowed them with the potential to be expressed through a continuous narrative. It became, in a way, the underlying organizational or grammatical matrix of the language of life.

Between body and mind: the spirit of language

The idea of a simultaneously material and metaphorical spirit traversing, ordering and rendering legible the world and the human perception of the world had been theorized by Montesquieu. His arguments, culminating in *De l'esprit des lois*, maintained that, just as the human body could adapt itself to fit a new climate through the reception of physical spirits (air, food etc.), so the human mind could modify thought through interaction with the outside world in the form of physical sensations and intellectual spirits such as ideas. The mind was shaped and changed through the circulation of these spirits. The surrounding environment, be it real or textual, influenced and altered the perceiving mind by a material, tangible process of osmosis: its spirits passed through the semi-permeable boundaries of the body, changing the economy, disposition and quantity of thoughts in the brain and causing the observer to see the world differently.

A description in *De l'esprit des lois* of an experiment in which Montesquieu examined a bit of tissue from a sheep's tongue under a microscope demonstrates how he thought this process of the movement of spirits worked. When he submitted the tissue to cold air, he noticed that the cells contracted and shrunk. As the cells defrosted, they became larger, taller and more spread out.[29] This activity, he reasoned, did not only happen on the microscopic level: like a domino effect, the same activity occurred on increasingly larger scales. When, for example, cold air touched 'les fibres' of the skin, they contracted and became more spring-like, charged with energy. They then imparted these qualities to the surrounding blood, though the vessels and all the way to the heart: 'cela [l'air froid] augmente leur ressort, et favorise le retour du sang des extrémités vers le cœur. Il diminue la longueur de ces mêmes fibres; il augmente donc encore par

28. Metzger, *Newton, Stahl, Boerhaave*, p.66 (original emphasis).
29. Montesquieu, *De l'esprit des lois*, p.476.

là leur force' ('Cold air contracts the extremities of the body's surface fibers; this increases their spring and favors the return of blood from the extremities [to] the heart. It shortens these same fibers; therefore, it increases their strength in this way too').[30] This chain of cause and effect was provoked by a change in temperature which subsequently affected the way fluids moved and interacted in the individual body. From these physiological observations, Montesquieu explained how temperature influenced not only physical fluids such as blood, but also less tangible 'spirits' such as passions, desires and intellect. The word 'esprit' thus referred simultaneously to 'thought', 'mind' and 'spirit' in both the abstract and the material senses.

The effect of a spirit could be equally well experienced by reading books as by experiencing nature. For example, in letter 143 of the *Lettres persanes*, Montesquieu established an analogy between books and medicine. Both, the letter claims, were able to change the proportion and kind of spirits in a body afflicted by illness, thus restoring a balance that was lacking and that had been causing poor health. A man with insomnia called upon a bookseller – and not a medical doctor – to recommend a multivolume work of theology to help him to fall asleep. When the book arrived, the son of the insomniac began to read it aloud, and the spirit of the author was directly transmitted to the child: he fell asleep, soon followed by the rest of the people in the room.[31] Upon hearing about the effect of this new remedy, the man's doctor decided to change his practices, declaring, 'Eh, pourquoi un esprit ne pourrait-il pas transmettre à son ouvrage les mêmes qualités qu'il a lui-même? [...] Je suis las des apothicaires: leurs sirops, leurs juleps, et toutes les drogues galeniques ruinent les malades et leur santé. Changeons de méthode; éprouvons la vertu des esprits' ('Ah, why cannot a mind transmit to its work the same qualities as it has itself? [...] I am tired of apothecaries: their syrups, their juleps, and all the Galenic drugs destroy the ill and their health. Let us change our method: let us test the virtue of spirits').[32] From this realization followed a list of prescriptions – not medicines,

30. Montesquieu, *De l'esprit des lois*, p.474; Montesquieu, *The Spirit of the laws*, p.231. In this English translation, the expression 'vers le cœur' is given as 'of the heart'. I do not agree with this, and have replaced 'of' with 'to' in the above text.

31. Montesquieu, *Lettres persanes*, in *Œuvres complètes*, ed. R. Caillois, vol.1, p.129-373 (356).

32. Montesquieu removed this part of the letter from the text, but, in a later edition of the *Lettres persanes* (1758), it appeared in a note. It can be found in the notes of the Caillois edition of Montesquieu, *Œuvres complètes*, vol.1, p.1599-1600.

but books – to be used for different kinds of ailments. This letter is of course ironic: on the surface it might be read as a mockery of both medicine and religion and their beliefs in the transmission of spirits. However, read in the larger context of the *Lettres persanes* as a reading program to change how people understood and interacted with their government – and in light of Montesquieu's later work in *De l'esprit des lois* – it is possible to interpret the letter as presenting an understanding of the idea of *esprit* as that which linked the metaphorical and the real, the mental and the natural or physical. Anne Vila's discussion of Charles Bonnet's 1760 *Essai analytique sur les facultés de l'âme* is a clear demonstration of how Montesquieu's theory of the relationship between sensitive bodies and language could be expressed in even more blatant materialist terms, merging them into one sensory and thinking organ. Vila writes that, for Bonnet, 'particular ideas are not merely associated with but also physically *attached* to particular fibers', allowing him to 'elaborate an extensive genealogy of sensibility – a genealogy that both explains the fibrillary derivation of all of man's higher faculties, and points the way to perfecting those faculties in the service of science, art, and morality'.[33]

Indeed, a language with *esprit* could be applied, Montesquieu argued, to the communication, appreciation and perpetuation of scientific knowledge and activities:

> Nous apprenons dans les livres de pur esprit l'art d'écrire, l'art de rendre nos idées, de les exprimer noblement, vivement, avec force, avec grâce, avec ordre et avec cette variété qui délasse l'esprit [...]. Les sciences se touchent les unes les autres; les plus abstraites aboutissent à celles qui le sont moins, et le corps des sciences tient tout entier aux belles-lettres. Or, les sciences gagnent beaucoup à être traitées d'une manière ingénieuse et délicate; c'est par là qu'on en ôte la sécheresse, qu'on prévient la lassitude, et qu'on les met à la portée de tous les esprits.

> In books of pure thought [or spirit] we learn the art of writing, of rendering ideas, of expressing them nobly and with life, force, grace, and with that order and variety that soothes the mind [...]. The sciences flow into one another; the most abstract ones lead into the less abstract, and the body of sciences relies entirely on *belles-lettres*. Now, sciences have much to gain from being handled in a delicate

33. Vila, *Enlightenment and pathology*, p.31-32 (original emphasis).

> and ingenious manner; in this way one avoids making them dry and weary, and renders them accessible to all minds.[34]

If works of natural knowledge were written by an author whose mind and senses were well trained in discerning and representing the order and movement of the world, the spirit of language could capture and communicate this world directly to the readers. The readers' minds would be saved from the 'dryness' and the 'weariness' that resulted from ignorance or from reading dull language that hinders the movement of thought and energy. This movement 'soothes the mind', refreshing it, causing it to reequilibrate and to come into harmony both with the world and with other productive minds.

An author capable of using language as a vessel to translate between world, text and reader had, according to Montesquieu, a mind which was unique, more refined and thus 'bien rare', in the sense of uncommon and esoteric, as well as in the sense of a rarefied, gaseous state of matter, which is less dense, more refined, and more elusive and difficult to imitate.[35] This rare and finely tuned quality of a genial mind was defined by 'delicacy' ('délicatesse'): a both more intense and more precise manner of experiencing the surrounding world, surpassing the usual limits of perception.[36] This delicate balance of spirits interacted with the environment, placing the author, and subsequently his or her public, both 'en état de juger' ('in a state to judge') and 'en état de plaire' ('in a state to please'). The delicate author '*sait* et il *sent* le juste rapport qui est entre les choses et lui. Un homme d'esprit *sent* ce que les autres ne font que *savoir*' ('*knows* and *feels* the well-founded relationship between him/herself and things. The man of spirit *feels* that which other authors can only but *know*').[37] The spirit that flowed from a text capable of bringing together sentiment and reason was meant to impart to its reader the essence of the subject itself, the language then able to appear more real and alive than the actual subject itself. In order for this understanding to take place, the physical receptors on the reader's body first had to be tempered, made less dry and more wet, less cold and more receptive to heat, so

34. Montesquieu, *Discours sur les motifs qui doivent nous encourager aux sciences*, in *Œuvres complètes*, ed. R. Caillois, vol.1, p.53-57 (56-57). This 1725 work was Montesquieu's reception speech to the Académie des sciences in Bordeaux.

35. Montesquieu, *Essai sur les causes qui peuvent affecter les esprits et les caractères*, in *Œuvres complètes*, ed. R. Caillois, vol.2, p.39-68 (58).

36. Montesquieu, *Essai sur les causes*, p.48.

37. Montesquieu, *Essai sur les causes*, p.57-58 (emphasis added).

that spirits could flow through them, to the brain. This movement, according to Montesquieu, would first be felt as a soothing revitalization, and only later interpreted as a heightened ability to know, to discover and then to judge and to act.

To the experience of the *esprit* of a text – or of the world – was added the critical, refining exercise of moderation and temperance. In the *Essai sur les causes qui peuvent affecter les esprits et les caractères*, Montesquieu wrote that: 'Nous devons être extrêmement ménagers des fibres de notre cerveau […]. [L]es mouvements modérés nous en promettent une infinité d'autres', while 'les violents prennent sur ceux qui doivent suivre' ('We must be extremely economical with the fibers of our brain[s] […]. [M]oderate movements promise us an infinity of others [movements]', while 'violent [movements] take over those that should follow').[38] Moderation, neither too hot nor too cold, increased the possibilities of both knowledge and creation of new modes of perception. 'Violent' thought, by contrast, merely followed the laws directing it, like a chain of action and reaction, leaving no opportunity to reflect upon another possible, and possibly better, path. The best writing not only showed readers the mechanism of the laws of nature, but also trained them in how to use their observations. They could observe something in nature and immediately abstract it, linking it to the deterministic, usual action of law. Alternatively, they could use their perception as an opportunity to see beyond the moment and beyond the pathways of action that nature had created, to new ones which involved the participation of the reader's reason and judgment. As later chapters in this book will show, Buffon eventually understood his language not only as a reenactment of nature's laws, and in particular the law of attraction, but furthermore as capable of transmitting to readers an advantageous perception of nature from which to visualize different analogies from above. They could learn to read nature as a great painting or machine – and then to reflect upon how this machine might be improved and made to function more smoothly.

The *Encyclopédie* article 'Chaleur' also showed how heat, because it changed perception, could be construed as simultaneously physical and poetic. In this article, D'Alembert described heat not as the measurement of temperature, but rather as the experience of a perceiving, feeling body. Heat was the sensation created by fire; it was 'une perception particulière ou une modification de notre âme' ('a

38. Montesquieu, *Essai sur les causes*, p.50.

unique perception or modification of our soul').[39] D'Alembert defined heat both as the movement and friction of particles in the body which was on fire and as the transfer of this movement to a receiving body. Usually, he explained, heat was evenly dispersed everywhere and thus undetectable. People were only able to perceive it and describe its presence in the moments when it escalated, causing a reaction and a physical change.[40] This exchange of movement triggered the experience of heat, light or color, depending not only upon how fast the particles moved, but also upon what state of matter the receiving body was inclined to sense. Experience was as important to the definition of heat as mechanics. Heat was 'une sensation relative', because it had to do with the communication of movement from one body to another such that the receiving body experienced a change in sensation, bringing the two bodies into dialogue.[41] Equal parts reality and perception, it could be thought of as an event: that is, both something that happened in the world *and* part of the story told of the world.

When D'Alembert's model of heat is combined with the rhetorical, discursive sense of heat seen in Montesquieu's work, the full power of this concept may be appreciated. Poetic language was understood to carry its message from author to reader through *energeia*, the realization of thought's potential in language. This activity was described, again by D'Alembert, in the *Encyclopédie* article 'Energie':

> ENERGIE, FORCE, synon. (*Gramm.*) Nous ne considérerons ici ces mots qu'en tant qu'ils s'appliquent au discours [...]. Il semble qu'*énergie* dit encore plus que *force*; & qu'*énergie* s'applique principalement aux discours qui peignent, & au caractere du style. On peut dire d'un orateur qu'il joint la *force* du raisonnement à l'*énergie* des expressions. On dit aussi *une peinture énergique, & des images fortes.*
>
> ENERGY, FORCE, synonyms. (*Grammar*) Here we only consider these words in terms of the application to discourse [...]. It seems that *energy* says even more than *force*; & that *energy* is applied primarily to discourses which paint, & to the character of style. One might

39. D'Alembert, 'Chaleur', in Diderot and D'Alembert, *Encyclopédie*, vol.3, p.23. D'Alembert linked his article on heat notably to the articles 'Sensation' and 'Perception'.
40. D'Alembert, 'Chaleur', p.23.
41. D'Alembert, 'Chaleur', p.23.

> say of an orator that he joins the *force* of reasoning with the *energy* of expressions. One also speaks of *an energetic painting, & forceful images.*[42]

This article was categorized under 'Grammar', and not under the physical sciences. An 'energetic' and 'forceful' spirit could be emitted by language, transmitting the power and order of the author's original ideas. As I will argue in the next chapter, the process of communicating both the structure and the vitality of the natural world was one of Buffon's main stylistic concerns in the *Histoire naturelle.* The 'character of style' in the above citation referred to a text's ability to 'paint' the nature of the author's thought – and, in the case of Buffon, the nature of the outside world as well.

Similarly, in his article 'Force' (also categorized under 'Grammar and literature'), Voltaire described how the meaning of this term 'a été transporté du simple au figuré' ('has been transported from simple to figurative'). From a mechanical definition of 'force' as 'toutes les parties du corps qui sont en mouvement' ('all the parts of a body that are in movement'), Voltaire then enumerated its metaphorical, analogical meanings, related to the expression of ideas through language – not demonstrative, mathematical language, which carried its own fixed degree of 'évidence', but eloquent, persuasive discourse that addressed 'les questions problématiques' ('questions of debate').[43] In a language similar to Buffon's *De l'art d'écrire*, Voltaire argued that '[l]a *force* de l'éloquence n'est pas seulement une suite de raisonnemens justes & vigoureux, qui subsisteroient avec la sécheresse; cette *force* demande de l'embonpoint, des images frappantes, des termes énergiques' ('The *force* of eloquence is not only in the succession of just & vigorous reasoning, which could subsist in coldness [or insensitivity]; this *force* demands embonpoint, striking images, energetic terms').[44]

The energy of the natural historical text

Throughout his career, Buffon's writings about nature consistently attempted to bring together a mechanical understanding of heat as the moving and changing of matter and its materialist and rhetorical senses as an energy or force. A combination of the rhetorical energy

42. D'Alembert, 'Energie, Force', in Diderot and D'Alembert, *Encyclopédie*, vol.5, p.651.
43. Voltaire, 'Force', in Diderot and D'Alembert, *Encyclopédie*, vol.7, p.109-10.
44. Voltaire, 'Force', p.110.

discussed by Voltaire with Buffon's contact with Newtonian physics and Hales' work on physiology and vital economy, as well as the likely impact of Montesquieu's theories about the circulation of spirits between bodies and the effects of their movement, provides a theoretical and technical foundation from which to understand his use of heat in the study of natural history.

Buffon considered his prose to be energetic, inspired *and* logical, scientific. The heat of the *Histoire naturelle*'s language was designed not only to put readers in awe of its author and the poetic worlds he created (for Buffon did indeed intend to show off his eloquent style), but furthermore to give them the epistemological and conceptual tools to experience the world through the experience of reading the text. The 'energetic strokes' described in *De l'art d'écrire* combined knowledge and inspiration, merging two terms of Aristotelian rhetoric: *enargeia*, clear and ordered knowledge, and *energeia*, action and movement, the realization of potential.[45] Through its vibrant, evocative discourse, the text of the *Histoire naturelle* communicated a systematic view of nature, one readers could use to better grasp the operation of the natural machine (and eventually learn to imitate and appropriate it). The vivacity of figurative, descriptive language that captured the activity of the world was counterbalanced by a measured clarity which prevented the structure, order and relationship of the historian's original ideas from becoming lost.[46]

Therefore, writing with heat did not necessarily mean trying to sway readers only by moving their emotions. Recalling Montesquieu's discussion of 'délicatesse', an effective author could transmit both feeling and knowledge through language. This might be read in contrast to the Enlightenment idea of genius as a poet who felt and expressed the world with grand strokes of the pen, operating from an innate base of uncontrollable talent – an idea that does not quite capture how Buffon thought of his work. In the article 'Génie' in the *Encyclopédie*, those who possessed this natural talent were distinguished from the *philosophes.* The philosophical thought process 'ne s'[accorde] guère avec la chaleur

45. Peter France, 'Lumières, politesse, énergie', in *L'Histoire de la rhétorique dans l'Europe moderne, 1450-1950*, ed. Marc Fumaroli (Paris, 1999), p.945-99 (980). On the rhetorical concept of *enargeia* in Antiquity, see Perrine Galand-Hallyn, *Le Reflet des fleurs: description et métalangage poétique d'Homère à la Renaissance* (Geneva, 1994), p.36-49.
46. Michel Delon, *L'Idée d'énergie en France au tournant des Lumières (1770-1820)* (Paris, 1988), p.36-37.

de l'imagination' ('rarely agrees with the heat of the imagination'); the philosopher organizes and interprets observations and experiences according to the logical course of reason. The genius, however, 'est frappé de tout' ('is struck by everything'), inspired and 'subjugué par l'enthousiasme' ('subjugated by enthousiasm'). He depicts the world with an energy that permits him to see what others cannot: 'il jette sur la nature des coups-d'œil généraux & perce ses abîmes. Il recueille dans son sein des germes qui y entrent imperceptiblement, [...] il observe rapidement un grand espace, une multitude d'êtres' ('he throws general glances upon the world and pierces its depths. In his heart he collects the seeds that imperceptibly enter there, [...] he rapidly observes a great amount of space, a multitude of beings').[47]

Buffon, by contrast, strove to be creative in a different manner, combining the cold, intellectual mind of the philosopher with the emotional heat of genius. He seems to have thought of himself as capable of both seeing nature from above and 'pierc[ing] its depths', of using the force of his language to lift the veil of the visible world to discover its hidden, invisible mechanisms. Such was the case in his hypothesis of the 'moule intérieur', the internal form passed down through the generations, defining animals of the same species.[48] Instead of explaining why or how a species replicates over time, Buffon considered how best to represent and transmit the occult workings of nature through language. The rational *and* poetic voice of the natural historian was necessary in order to grasp the productive, dynamic and subtle activity of nature, and to render this activity legible, sensible, to readers. Buffon discerned and ordered the visible effects, what he called 'les moyens apparens', of belonging to a specific species (resemblance, ability to produce viable offspring). The ordered chain of effects then allowed him to 'imagine' an underlying natural process and a narrative flow, most famously communicated in his articles on the horse, the donkey and the dog.[49] Thus he sought to capture life, focusing on the generation and development

47. Jean François de Saint-Lambert, 'Génie', in Diderot and D'Alembert, *Encyclopédie*, vol.7, p.583.
48. On the *moule intérieur* and Buffon's theory of generation, see Jacques Roger, *Les Sciences de la vie dans la pensée française du XVIIIe siècle* (Paris, 1963), p.542-58. See also Reill, *Vitalizing nature in the Enlightenment*, p.47.
49. Buffon, 'De la reproduction en général', in *Histoire naturelle* (1749-1767), vol.2, p.126-27. On Buffon's definition and language of species, see Hanna Roman, 'Naming as natural process and historical narrative in Buffon's *Histoire naturelle*, 1749-55', *Romance studies* 31:3-4 (November 2013), p.238-50 (244-48).

of organisms over time such that members of the same species resembled one another yet constantly changed, interacting with and adapting to their environments. He studied and recorded both the alteration of external form and the ability to both grow and decay, to slip between states of more or less energy and life, to transgress boundaries and develop new identities as part of an ongoing dialogue with the surrounding world.

How to convey this invisible, underlying force that made nature alive? Recalling the idea of 'painting with words' from *De l'art d'écrire*, Buffon argued the best way was through the energy of a language that literally vibrated on the same frequency as the world it described. This language should seek to harmonize with the movement of the subject matter, and not try to embellish it with the artifices of poetry or painting. In a manuscript page entitled 'Notes sur la façon de dessiner un cheval' ('Notes on the way to draw a horse'), from a collection of notes relating to the composition of the *Histoire naturelle*, Buffon contrasted the representation of a perfectly proportioned horse with how a real horse would look and act. Breeders imagined producing a horse according to an ideal model in accordance with 'les proportions de la belle nature' ('the proportions of beautiful nature'), in much the same way that artists sought an ideal model in human forms immortalized in sculpture, 'comme celles de l'Apollon du Vatican et de la Venus de Medicis' ('like […] the Apollo at the Vatican or the Venus of the Medicis').[50] When represented, such attempts to perfect nature degenerated into monstrous composites of reality and imagination, artificial creations that neither embodied nor transmitted their own life and energy:

> [I]l n'y a aucun homme [,] aucune femme ni aucun cheval qui ait les proportions de la belle nature […]. Un individu qui auroit toutes ces proportions est donc un être imaginaire, par conséquent la belle nature n'est pas la vraie nature qui est l'objet du naturaliste […]. Un dessinateur qui se propose de représenter un bel homme, réunit dans la figure, qu'il en fait, toutes les proportions de la belle nature: cette figure n'existe dans aucun individu de l'espèce humaine; c'est un produit de l'art […]. Dessiner d'après nature, suivant l'expression de l'art, c'est représenter fidelement l'objet, que l'on a sous les yeux, sans y rien changer par addition, par suppression, ni d'aucune autre manière.

50. 'Notes ayant servi pour la rédaction des œuvres de Buffon', Paris, Bibliothèque du Muséum national d'histoire naturelle, MS 218, f.3-4.

> [T]here is no man, woman or horse who has the proportions of beautiful nature […]. An individual with all these proportions would be an imaginary being, and thus beautiful nature is not the same thing as true nature, which is the object of the naturalist […]. A draftsman intending to represent a beautiful man brings together in the face which he draws all of the proportions of beautiful nature: that face does not exist in any individual of the human species; it is a product of art […]. Drawing from nature, following the manner of art, faithfully represents the object being observed, without adding to it, deleting from it or changing it in any other way.[51]

In his essay on horses in the *Histoire naturelle*, Buffon also applied the argument that writing, and not drawing or painting, was the only way to 'dessiner d'après nature' ('to draw from nature'). A drawing revealed a predetermined explanation of the world, designed to fit a certain model or system, showing only how people thought. It was not receptive to the real operation of nature. Buffon inverted the idea of 'la belle nature' as created by poetry. He claimed that on its own 'la Nature est plus belle que l'art, & dans un être animé la liberté des mouvemens fait la belle Nature' ('Nature is more beautiful than art, and in an animated being the liberty of movements creates beautiful Nature').[52] And yet, while doubting the ability of art to render nature more beautiful, Buffon wanted his prose to convey the world, by discovering and emphasizing nature's inherent poetry, its underlying creative processes. This hidden movement of nature could be felt or experienced through continuous observation over time, revealing the measured relations – the poetry – among phenomena. Buffon intended to mirror this activity in his prose, thus elevating prose to the status of poetry, or perhaps surpassing poetry altogether. The movement of his words, the manner in which they flowed, could give vitality to his ideas, the same vitality observed in the natural world. Yet, the desired effect of his language was not to be fanciful, charming or awe-inspiring, but to use the lyrical channels of discourse to discover

51. 'Notes ayant servi', f.3-4.
52. Buffon, *Histoire naturelle* (1749-1767), vol.4, p.176. For Buffon, systems, definitions and classifications were decorative, superficial and even deceptive. He argued, for example, that by dividing animals into six classes Carolus Linnaeus' work stated what was in nature, but failed to show how nature worked. For Buffon's critique of Linnaeus and methods of classification in natural history in general, see the 'Premier discours' of the *Histoire naturelle* (1749-1767), vol.1, p.13-20 and 37-41. See also Phillip R. Sloan, 'The Buffon–Linnaeus controversy', *Isis* 67:3 (September 1976), p.356-75.

the underlying energy of nature, and thus to transmit to the rational mind the order and operations of the world.

Having raised prose to the status of poetry, Buffon then inverted the concept of *ut pictura poesis*, claiming not that poetry – now understood as expressive prose – resembled painting, but rather that painting was a form of poetry.[53] Moreover, he maintained, painting was best realized through the temporal, life-like unfolding of events. Only the act of painting could accurately and authentically depict an object – better, it would seem, than nature itself – and painting was best done with words. Not just any words, but the consecutive, flowing succession of energetic language on the page: prose.[54] What was being rendered in the mind of the reader was not a picture but a chain of words, a carefully linked set of relationships. Only such an *enchaînement* of words could encapsulate and reproduce the dynamic flow of nature through time, inserting Buffon and his readers within this flow.[55] In Michel Delon's words, '[l]'énergie devient une catégorie fondamentale du jugement littéraire et assure une prééminence de la prose' ('[e]nergy becomes a fundamental category of literary judgment and insures the preeminence of prose').[56] Once the heat of nature was truly felt and understood, the science of natural history could become useful. Its language could be used to interpret, refine and redirect the potential of heat.

For Buffon, the ability of language to transmit the experience of seeing a horse move, for instance, was produced by the natural historian through constant observation of horses over time – in the same manner as he observed and depicted species over time. Observation was a process through which the author of a natural history tracked the development of a phenomenon, relating its present state in the surrounding world

53. In the Enlightenment a common debate around *ut pictura poesis* was a competition between artistic and scientific disciplines about which could better capture the essence of nature. See Liebman, 'Painting natures', p.16.

54. 'Buffon changed the terms of the eighteenth-century debate on the relationship between poetry and painting. In particular, he championed prose over poetry as a superior vehicle for visual representation, claiming that verse curtailed the writer's ability to paint [...]. To the extent that it imposed no artificial constraints on the writer, prose was more permeable to the qualities of nature. It allowed the writer to develop a natural style, which reflected the actual progression of his thoughts and perceptions rather than arbitrary formal rules.' Stalnaker, *The Unfinished Enlightenment*, p.64.

55. Wilda C. Anderson, *Diderot's dream* (Baltimore, MD, 1990), p.40.

56. Delon, *L'Idée d'énergie en France*, p.131.

to its past and its future. Observation was about nuancing vision and learning to see past the visual to uncover the poetic structure and rhythm of nature through history.[57] The movement of a horse was not simply the mechanical displacement or the physical activity of its muscles and bones: the movement of species through history, and the development of a relationship between a species and its climate, or between a species and people, also had to be considered. The heat of language about horses had to combine corporeal, local movement with a more universal, unseen – but equally vital – movement.

Buffon's article on the horse presents two very striking examples of what he meant by giving movement and life to a subject. The first and most famous citation, from the very beginning of the article, depicts the horse as it has been shaped by humans:

> La plus noble conquête que l'homme ait jamais faite est celle de ce fier et fougueux animal qui partage avec lui les fatigues de la guerre et la gloire des combats; aussi intrépide que son maître, le cheval voit le péril et l'affronte, il se fait au bruit des armes, il l'aime, il le cherche et s'anime de la même ardeur; il partage aussi ses plaisirs, à la chasse, aux tournois, à la course, il brille, il étincelle; mais docile autant que courageux, il ne se laisse point emporter à son feu, il sait réprimer ses mouvemens, non seulement il fléchit sous la main de celui qui le guide, mais il semble consulter ses desirs, et obéissant toûjours aux impressions qu'il en reçoit, il se précipite, se modère ou s'arrête, et n'agit que pour y satisfaire; c'est une créature qui renonce à son être pour n'exister que par la volonté d'un autre, qui sait même la prévenir, qui par la promptitude et la précision de ses mouvemens l'exprime et l'exécute, qui sent autant qu'on le desire, et ne rend qu'autant qu'on veut, qui se livrant sans réserve, ne se refuse à rien, sert de toutes ses forces, s'excède et même meurt pour mieux obéir.
>
> The noblest conquest that man ever made is that of this proud and spirited animal, who shares with him the fatigues of war and the glory of victory. As daring as his master, the horse sees danger and confronts it; he is animated by the noise of weapons, he loves it, he seeks it and

57. 'What we think of as a simple process of visual perception is [...] actually a complex mental process combining several forms of sensory perception [...]. By incorporating the full range of senses into his text, the painting writer brings the reader closer to his own experience of first-hand observation. In doing so, he infuses his text and his reader with the warmth and life of the natural world.' See Joanna Stalnaker, 'Painting life, describing death: problems of representation and style in the *Histoire naturelle*', *Studies in eighteenth-century culture* 32 (2003), p.193-227 (218).

> comes to life with the same ardor [as his master]. He also shares [his master's] pleasures in hunting, in tournaments, in races, he shines, he sparkles; but just as docile as he is courageous, he does not give in to rash behavior, he knows how to restrain his movements. Not only does he yield to the hand of his guide, but he seems to consult his [guide's] desires, and always obeying the impressions he receives, he speeds up, controls himself or stops, acting only to satisfy [the master's desires]. The horse is a creature who gives up his being to exist only for the will of another; who can anticipate it [this will] in advance; who, by the promptitude and precision of his movements, expresses and executes it; who feels as much as one wants him to and only gives as much as one wants; who, by giving without reservation, refuses nothing, uses all his forces, exceeds himself, and even dies to better obey.[58]

The rhythm and stresses of the first clause of this passage, when read aloud in French, sound like the gallop of a horse. It could almost be separated from the rest of the paragraph, put into verse, and read like a traditional poem. It has the strict and controlled meter of 'beautiful nature', for it is indeed an imitation of people's manipulation of nature, shaping the creatures around them to accommodate their needs and practices, codeveloping the species of the horse with nature. The horse painted here is artificial, not natural, and thus the movement of this passage is rhythmic and artificial. Nevertheless, Buffon's style here is natural to, and hence in harmony with, its human-made subject matter.

The next example directly follows the previous citation in the essay 'Le cheval', and reveals how the flow of Buffon's ideas changed when talking about a different kind of movement: of the horse in the wild, unaltered by human art and practice.

> La Nature est plus belle que l'art, et dans un être animé la liberté des mouvemens fait la belle Nature: voyez ces chevaux qui se sont multipliés dans les contrées de l'Amérique Espagnole, et qui y vivent en chevaux libres, leur démarche, leur course, leurs sauts, ne sont ni gênés ni mesurés; fiers de leur indépendance, ils fuient la présence de l'homme, ils dédaignent ses soins, ils cherchent et trouvent eux-mêmes la nourriture qui leur convient, ils errent, ils bondissent en liberté dans des prairies immenses, où ils cueillent les productions nouvelles d'un printemps toûjours nouveau; sans habitation fixe, sans autre abri que celui d'un ciel serein, ils respirent un air plus pur que celui de ces Palais voûtés où nous les renfermons en pressant les espaces qu'ils doivent occuper; aussi ces chevaux sauvages sont-ils beaucoup

58. Buffon, *Histoire naturelle* (1749-1767), vol.4, p.174.

> plus forts, plus légers, plus nerveux que la pluspart des chevaux domestiques, ils ont ce que donne la Nature, la force et la noblesse, les autres n'ont que ce que l'art peut donner, l'adresse et l'agrément.

> Nature is more beautiful than art, and in an animated being, the liberty of movement creates beautiful nature: consider those horses who reproduced in the Spanish regions of the American continent, and who live there as free horses. Their gait, their run, their jumps are neither awkward nor measured; proud of their independence, they flee the presence of man, they scorn his care, they look for and find for themselves the food that they need, they wander, they leap in liberty in immense prairies where they gather the new growths of an eternally new spring. Without a fixed dwelling, with no other shelter than that of a serene sky, they breathe a purer air than that of those vaulted palaces where we enclose them, squeezing them into the spaces they must occupy. These wild horses are also stronger, lighter, more nervous than the majority of domesticated horses. They have what nature gives them: force and nobility. The others have only what art can give: skill and charm.[59]

Depicting wild horses' life and movement with respect to their surroundings, the prose of this passage does not have the same regular rhythm as the consistent beat of the tamed horse's gallop. Through the less tightly controlled tempo of its language, this passage paints wild horses as 'plus forts, plus légers, plus nerveux' than domesticated ones, for 'leur demarche, leur course, leurs sauts, ne sont ni gênés ni mesurés'. Both Buffon's language and the wild horses 'ont ce que donne la Nature, la force et la noblesse'. Domesticated animals, like texts that follow the conventional rules of poetry, lack this ability to transmit the energy of nature, and 'n'ont que ce que l'art peut donner, l'adresse et l'agrément'.

Through the depiction of movement, written discourse transmitted the identity and meaning of an object, as well as its relationship to the larger structure of nature, in a subtler fashion than a static image. From the physical movement of the horse, wild or tame, the more hidden, historical movements of the development of its species and its relationships to people could be felt. The two passages about horses show how Buffon's prose was, in Stalnaker's words, 'more permeable to the natural world'.[60] It was the medium between the energy of the world and the mind of the reader. The author had to use all of the

59. Buffon, *Histoire naturelle* (1749-1767), vol.4, p.175-76.

60. Stalnaker, 'Painting life', p.221.

physical senses ('l'emploi de tous les sens', to recall the first citation from *De l'art d'écrire*) in order to transmit a dynamism not always visible but certainly palpable, in the form of an energetic warmth that reanimated the natural world in the reader's mind. Thus heat, at this stage in the *Histoire naturelle*, was not really the measurement of temperature as much as it was the invention of a new way of perceiving the world. Reading the *Histoire naturelle* was meant to set the reader's mind ablaze in a way that could not have happened either by directly observing nature or by reading an ordered description or *tableau* of the world. The reader was to be left with neither a momentary glimpse of the world nor a fixed, rigid structure, but with a lasting heat, the continuation of nature's ability, capacity and history, through the natural historical text, to the mind.

Buffon's language taught readers to become differently sensitive to the world so that they viewed it no longer as a stagnant collection of 'dead and detached parts' (to refer back to *De l'art d'écrire*), but rather as a fluid and dynamic network of relationships that connected the vitality of nature to the energy of Buffon's prose to the receptivity of the reader's mind. The language of natural history was carefully regulated; it did not shock with sudden blasts of heat or sparks, but captured and manipulated the energy of nature to produce a discourse at once powerful and ordered. The heat of language imparted the sequence of natural events, as well as their dynamism and the changes they sparked in both the course of history and the perception and ability of humankind. The result of this activity was the creation of a new, textual nature which could not only inspire readers but be useful to them.[61] Furthermore, natural history not only related the events that had taken place, but conjectured upon the future of the world (an idea analyzed in Chapters 4 and 5). By modifying the perceptive ability of readers, the heat of language opened up the potential for creative change and innovation in the world.[62]

61. This is similar to how Anderson describes Diderot's definition of a genius as '[t]he person capable of seeing through to the hidden unity and working with it [...]. Diderot distinguished between the local genius as the intense talent for a particular discipline [...] and comprehensive genius with an uncanny ability to relate one's work to the rest of nature's functioning. The creator sees beyond. As a result he innovates rather than just discovering; this is what it means to have his work characterized as *utile*.' Anderson, *Diderot's dream*, p.19.
62. Reill, *Vitalizing nature in the Enlightenment*, p.89-90.

3. Writing nature: the foundations of natural history

> Truth is not a convention, *but rather (1) in historical perspective, an event in the history of thought, (2) in its contemporary context, stylized thought constraint.*[1]

In the study of literature, Buffon is best remembered today for his 'Discours sur le style'. This work is the key to understanding Buffon's language as a form of natural knowledge, generated through observation and its translation into writing, that is, lyric, energetic discourse. The 'Discours sur le style' reveals the importance of first studying the *Histoire naturelle* as a literary object, in order to access, absorb and use its knowledge.

As I suggested in the introduction, style is a multifaceted term and its definition is not always obvious. The 'Discours sur le style' is often studied alongside Enlightenment literature to demonstrate the meaning of style by justifying the appearance at the time of a more modern, recognizable sense of authorship: an individual whose writing reflects not only the esthetic rules of the day but also his or her own character and unique voice. According to this twofold definition, style may refer either to a general set of rules and practices or to an individual's approach and to what degree it resembles or deviates from the general rules. In its etymology, according to Anna Wessely,

> the Latin word *stilus* (writing instrument) directed attention to the particular individual features of representations; the Greek word *stylos* (column) referred to the different orders of architecture that were, in turn, associated with the public significance of buildings. Their

1. Ludwik Fleck, *Genesis and development of a scientific fact*, translated by Fred Bradley and Thaddeus J. Trenn, ed. Thaddeus J. Trenn and Robert K. Merton (Chicago, IL, 1979), p.100 (original emphasis).

> connotations also differed: *stylos* provided an expressly social or moral foundation for aesthetic judgments; *stilus* stressed the shrewdness of observations, the ability and facility with which they were recorded – – in short, it emphasized the cognitive and technical basis of representations.[2]

The idea of style in works of science is borrowed from art history and literature. In scientific practice today, the scientific method could be considered to be the equivalent of the *stylos* of architecture, connoting universality and sameness, providing a global model to follow. Yet it is important to underline that style is a historically derived term that has held different meanings over time. The more nuanced, individual *stilus* is also found in the study of the history of science, in the recognition of different cultural practices and ways of reading and interpreting the world. Paradigms and subparadigms develop, interact and change over time, according to Jean Gayon, each with its own method, language and relationship between these two.[3] In this way, style was and continues to be the underlying base of epistemological and ontological assumptions in a specific culture, defining how meaning is made and organized. Wessely thus describes style as a 'system of meanings within the closed construction of a world. It determines what can be regarded as a solid fact or an acceptable explanation.'[4] It prevents seeing scientific knowledge as reductive or universal, showing instead 'the tacit and culturally bound rules that govern scientific research', the intellectual processes by which knowledge is formulated.[5] Indeed, for many natural philosophers of the 1750s, researching the world meant discovering this hybrid zone between knowledge and the literary, discursive tools of its creation.

A *stylus* was also a 'writing instrument', a tool which rendered ideas visible and gave them form. As Claudie Balavoine has written, through style it is possible to think about words, 'la matière verbale', like moldable clay in which will be left the mark or the sign of an idea

2. Anna Wessely, 'Transposing "style" from the history of art to the history of science', *Science in context* 4 (1991), p.265-78 (266).
3. Jean Gayon, 'De la catégorie de style en histoire des sciences', *Alliage* 26 (1996), n.p.
4. Wessely, 'Transposing "style"', p.273. Here, Wessely draws from Ludwik Fleck's construal of how facts and meaning were produced through relationships between groups of scientists studying syphilis and the concept of disease. See Fleck, *Genesis and development of a scientific fact.*
5. Wessely, 'Transposing "style"', p.276.

or a thing. In the original Greek, she explains, 'character' referred to 'toute marque en relief dont on se sert pour imprimer une figure dans une matière' ('every mark in relief which one uses to impress a figure into matter'); in this way, 'le discours sera [...] comparé à la matière malléable susceptible de recevoir une empreinte prédéterminée' ('the discourse will be [...] comparable to malleable matter susceptible to receiving a predetermined mark'). Style was the linking of word and thing through a figure or sign; 'il permet à la *res* imaginaire [...] de prendre corps' ('it allows the imaginary thing [*res*] [...] to become corporeal'). According to Balavoine's article, the first goal of words was not to create pleasure or beauty, but to transmit reality, to produce the experience of mimesis.[6]

For Buffon, style was both *stylos* and *stilus*: a method of presenting, ordering and explaining nature *and* a mode of rendering the natural world real or corporeal. According to his 'Discours sur le style', in its initial stages style was a method. It was the set of rules, as discussed in Chapter 1, through which an observer learned to measure his or her thought with respect to the events and laws of the outside world, bringing order to ideas. Heat was the mode through which the ideas were made tangible, fusing the activities of mind with those of the world. Through the combination of method and mode of representation, the natural historian channeled the energy and vitality of the world, and translated its objects into written figures on the page, harnessing what Ernst Cassirer has called the 'creative power from which the form and the order of the universe are derived'.[7]

The purpose of style for Buffon was to embody natural history: both the interactions between natural objects and phenomena and the movement and processes of nature through time. He argued that language could eventually come to do the same thing as nature, and in the same manner: that the words and ideas would follow one another across the pages just as the creations and events of nature succeed one another across time. In the *Histoire naturelle*, the naturalization of

6. Claudie Balavoine, 'La *Poétique* de J.-C. Scaliger: pour une mimésis de l'imaginaire', in *La Statue et l'empreinte: la poétique de Scaliger*, ed. C. Balavoine et P. Laurens (Paris, 1986), p.107-29 (118-19). On style in literature, see also Antoine Compagnon, *Le Démon de la théorie: littérature et sens commun* (Paris, 1998), ch.5 ('Le style'). Another interesting reflection on style, including Buffon's use of the term, may be found in David Bellos, *Is that a fish in your ear? Translation and the meaning of everything* (New York, 2011), ch.26 ('Style and translation').
7. Ernst Cassirer, *The Philosophy of the Enlightenment*, translated by Fritz C. A. Koelln and James P. Pettegrove (Princeton, NJ, 1951) p.326.

the historian's thought to nature, or the poetic activity through which he believed he could render the relational structure of his ideas and words analogous to that in nature, was a temporal process. Through the process of style, Buffon's conception of history can be grasped.

The 'Discours sur le style': translating the movement of nature

Every text, Buffon wrote in the 'Discours sur le style', had an underlying 'plan' upon which observations of the subject matter were translated into ideas, arranged and then articulated. He echoed the familiar trope of nature as a book: the world, too, had a plan, an underlying blueprint for its actions. Buffon's language depicted nature as a narrative that might be read like a book: a 'work' with a 'plan' and a chronological movement. To write well, Buffon stated, one should try to follow the productive process laid out in this quintessential plan, which contained all the conditions and instructions for nature's 'works':

> Pourquoi les ouvrages de la Nature sont-ils si parfaits? c'est que chaque ouvrage est un tout, et qu'elle travaille sur un plan éternel dont elle ne s'écarte jamais; elle prépare en silence les germes de ses productions; elle ébauche par un acte unique la forme primitive de tout être vivant: elle la développe, elle la perfectionne par un mouvement continu et dans un temps prescrit.

> Why are the works of Nature so perfect? It is because each work is a whole, and because she [Nature] works upon an eternal plan from which she never deviates. Silently, she prepares the seeds of her productions; in one action she sketches the primitive form of every living being: she develops it, perfecting it through a continuous movement and in a prescribed time.[8]

Buffon created the effect of a seamless continuity between the text (the work of the historian) and the world (the work of natural law). Enabled and structured by the operation of natural laws over time, nature appeared to embody the perfectly composed text. A writer would want to be immersed in nature, the ideal text, in order to compose a story of the world that unfolded with the same flowing and regular motion as nature itself, and to develop ideas into language in the same ordered

8. Buffon, 'Discours sur le style', p.5-6.

and organic manner in which nature transformed its 'seeds' into physical objects. The 'Discours sur le style' opened with a persuasive rhetoric on writing a text with style, showing the result of conflating ideas in the mind with objects in the world, such that the text, the ordered body of language, had in its very essence the potential to resemble and represent the productions of nature.

In the above quotation, the 'plan éternel dont elle ne s'écarte jamais' was the essence of nature's 'productions': everything observed in the world was a result of this fundamental and universal structure. In the plan, the story of an entire world might be imagined and generated, for it contained the instructions for writing like nature. Nature was depicted as an author: guided by laws, it sketched ('elle ébauche'), as if through a single gesture ('par un acte unique'), the basic draft of every organism. From this 'primitive form', nature built and refined its productions 'through a continuous movement and in a prescribed time' ('par un mouvement continu et dans un temps prescrit'), subjecting each aspect of its creation to the same conditions of law and time, providing it with its present form. The natural historian's method should mirror the continual process of natural creation, with a language like an uninterrupted brushstroke, perfected within a necessary period of time. The goal was to develop a poetic power that possessed the same ability, and operated in the same manner, as the creative power of natural laws – except through words and not the things or events themselves.

Style was the result of reproducing natural activity through observing and reading the world over time. Although not directly stated in the 'Discours sur le style', Buffon's conception of style implied that time was the medium and the space in which natural laws created and unfolded their 'plan', the realization of nature's 'seeds'. Time was the canvas on which nature's history was painted. In the 'Premier discours' of the *Histoire naturelle*, Buffon showed how the work of the natural historian, too, was constructed through time: as he became more experienced and meticulous in his skills of perception and judgment, initial observations of individual beings were linked to subsequent ones, forming ever more general understandings of nature over both space and time. The result of style, Buffon suggested, was that, just as he read the book of nature, the writer of history would eventually be able to read back through the history of his rational activity, perceiving it as a chain of cause and effect leading both logically and chronologically back to the original observations. The written text came to embody both the story of nature and the record of the historian's experience of understanding it.

The history of the text began before ever being written down. If style, as Buffon defined it, 'n'est que l'ordre et le mouvement qu'on met dans ses pensées' ('is nothing but the order and movement that one puts in one's thoughts'), the writer had first to understand how to create order and movement. Before ordering his perceptions of the world on the page, he had to discover the 'premières vues' and 'principales idées' from which they came, that is, the origins from which the rest of the work would subsequently flow.[9] In this Platonic primary form lay the seed of the 'premier plan', from whence the subject of the text would germinate, take on substance, movement and order in the form of language. Unearthing the essential idea, Buffon explained, was the first step in generating style:

> Mais avant de chercher l'ordre dans lequel on présentera ses pensées, il faut s'en être fait un autre plus général et plus fixe, où ne doivent entrer que les premières vues et les principales idées: c'est en marquant leur place sur ce premier plan qu'un sujet sera circonscrit, et que l'on en connoîtra l'étendue; c'est en se rappelant sans cesse ces premiers linéamens, qu'on déterminera les justes intervalles qui séparent les idées principales, et qu'il naîtra des idées accessoires et moyennes qui serviront à les remplir [...]. Pour peu que le sujet soit vaste ou compliqué, il est bien rare qu'on puisse l'embrasser d'un coup-d'œil, ou le pénétrer en entier d'un seul et premier effort de génie; et il est rare encore qu'après bien des réflexions on en saisisse tous les rapports. On ne peut donc trop s'en occuper; c'est même le seul moyen d'affermir, d'étendre et d'élever ses pensées: plus on leur donnera de substance et de force par la méditation, plus il sera facile ensuite de les réaliser par l'expression.

> But before seeking the order in which one will present one's thoughts, it is necessary to have made another order, more general and more fixed, in which can only enter the primary views and the principal ideas. By marking their place in this initial plan, a subject will be defined and its scope will be known; by constantly recalling these primary features one will determine the correct intervals which separate the principal ideas, and the accessory and middle ideas that will fill the intervals will be born [...]. Providing that a subject is vast or complicated, it is quite rare that one can comprehend it in the blink of an eye, or understand it entirely in a single, initial stroke of genius;

9. Buffon, 'Discours sur le style', p.3-5.

> and it is even more rare that after many reflections one can grasp all the relationship. One cannot but pay too much attention to it; this is indeed the only way to affirm, to expand, to lift one's thoughts, and the more one gives them substance and force through meditation, the easier it will then be to make them real though expression.[10]

One could not immediately grasp the entirety of nature. Understanding the world meant to work as it did, through time, in order to discover nature's essence, arrange and relate observations and perceptions, and make them real and visible through language ('les réaliser par l'expression'). The natural historian could not 'but pay too much attention' to this task. The beginning of the naturalist's story was not the same as the beginning of the natural world (a complex theological and epistemological problem that Buffon confronted in his hypothesis on the origins of the world, discussed in the next chapter). The *Histoire naturelle* did not originate in the potential of natural law, but rather in the unique nature of the writer's mind, which Buffon thought should possess an innate philosophical and genius-like capacity for reason, for comparing and ordering ideas and for translating these ideas into a cohesive story and rigorous argument. A writer capable of producing style would begin the story – of the subject and of the experience of studying it – from the potential of his or her own natural and rational ability.

The result of reason was what Buffon called 'une suite, une chaîne continue' ('a succession, a continuous chain'), a temporal sequence of thoughts in which each subsequent thought became dependent upon the others before it. His writing would then reflect this logical and temporal movement:

> Pour bien écrire, il faut donc posséder pleinement son sujet, il faut y réfléchir assez pour voir clairement l'ordre de ses pensées, et en former une suite, une chaîne continue, dont chaque point représente une idée; et lorsqu'on aura pris la plume, il faudra la conduire successivement sur ce premier trait, sans lui permettre de s'en écarter, sans l'appuyer trop inégalement, sans lui donner d'autre mouvement que celui qui sera déterminé par l'espace qu'elle doit parcourir.

> To write well, it is therefore necessary to fully possess one's subject, to reflect on it enough to clearly see the order of one's thoughts, and to form a succession of thoughts, a continuous chain, of which each link represents an idea; and once one has taken up the pen, it will be

10. Buffon, 'Discours sur le style', p.3-4.

> necessary to guide it successively from this first line, without letting it stray, without pressing it too unevenly, without giving it any other movement than that which will be determined by the space that it must cross.[11]

While events in nature and the historian's account of them were not the same thing, as the historian persisted in the study of nature's activities, the 'Discours sur le style' implied that the order and procession of the text of natural history would gradually transform into a convincing reproduction of nature's own work. The work of the natural historian and the work of nature eventually grew to be inseparable; the writer learned true expression through the investigation of natural history and the repetition of its processes in ideas and language. In fact, Buffon thought that without this tuning or harmonization of the mind to the natural world, the human spirit alone lacked enough inventive power:

> L'esprit humain ne peut rien créer, il ne produira qu'après avoir été fécondé par l'expérience et la méditation; ses connoissances sont les germes de ses productions: mais s'il imite la Nature dans sa marche et dans son travail, s'il s'élève par la contemplation aux vérités les plus sublimes, s'il les réunit, s'il les enchaîne, s'il en forme un tout, un système par la réflexion, il établira sur des fondemens inébranlables, des monumens immortels.

> The human mind cannot create anything. It will not produce anything until after having been fertilized by experience and meditation; its knowledge is the seed of its productions. But if it imitates Nature in its march and in its work, if it raises itself through contemplation to the most sublime truths, if it reunites these truths, if it links them, if it forms them into a whole, a system made by reflection, it will establish immortal monuments upon unshakable foundations.[12]

By suggesting that the human mind was most productive when imitating the creative movement of nature, and that the text with style was a result of being inspired by nature's 'most sublime truths', the 'Discours sur le style' might be read not so much as a general guide for how to write as a specific program for writing natural history. It showed just how closely the development of style was linked to the study of nature. Human reason alone, even with its ability to reflect

11. Buffon, 'Discours sur le style', p.8.
12. Buffon, 'Discours sur le style', p.6.

upon experience, was not enough to write a text that had style. People could not *create*, they could only *produce*: that is, they could construct from preexisting materials, but they could not bring something into existence from nothing. Nature alone created, in a single act of force, while the writer, 'fertilized' ('fécondé') by observations of nature's work, imitated its eternal gesture and its primordial idea, thus deriving the productive energy necessary to make the seeds of his or her own thoughts grow.

Yet it was precisely through this imitation, allowing himself to become the scribe who translated nature's 'marche' and 'travail' into language, that the historian began to invent a new model of perception by modifying and perfecting the natural, innate character of his mind. As he studied nature's activity, he not only refined his observations, but learned to make his account work like nature, producing through language in the same manner as nature created through its power. Nature gradually shaped the world from the created seed; so, too, the historian progressively drew analogies between observations, and eventually learned to see the world on larger and larger scales of perception, moving toward the 'most sublime truths' ('les vérités les plus sublimes'), the hidden mechanisms of nature.

After elevating reason to the level of nature's work, after generating a system that allowed both the force or heat of nature and the processes of natural history to be seen from above, the greatest act of the natural historian was, according to Buffon, the formation of 'immortal monuments' ('des monumens immortels'). Monuments were first and foremost the result of humankind's production: in the above citation, Buffon stated that the human mind 'will *establish* them ('il [les] *établira*'; my emphasis). They were to be built upon 'unshakable foundations' ('des fondemens inébranlables'), that is, upon the historian's underlying plan, formed through the collaboration of the activity of nature with that of natural history. Yet monuments were neither the one nor the other, but rather the interface between nature's processes and the natural historian's writing. They were, as such, the material proof of interaction between people and nature. Buffon's text, his style, was thus also and especially a monument – at once philosophical, historical and discursive – the language or physical residue of the process of coming into harmony with nature. The natural historical text possessed style not only because it was the story of nature produced through the mouthpiece of the natural historian, but indeed because the natural historian learned to use the work of nature to become a coauthor of history. The style of the text would,

Buffon believed, allow it to be understandable to future generations of readers, who would be able to reconstruct and relive the experience of learning about nature.

The *mise-en-scène* of style in the *Histoire naturelle*, 1749

In the 'Premier discours', the introductory essay to the first volume of the *Histoire naturelle*, Buffon began to stage the work of style by analyzing the process through which the vast and varied natural world might be transformed into both a science and a text. As Descartes introduced the *Discours de la méthode*, Buffon claimed to begin his research with no prior assumptions, criticizing the methods and systems of other natural philosophers. Unlike Descartes, he asked the observer to plunge into the apparent confusion of the physical, outside world and to start making regular observations.[13] At the beginning of the 'Premier discours', observers of nature were depicted as outsiders looking in at a great array of things of which they were yet unable to make sense. If a universal and true order of nature existed, they were not capable of perceiving it or of seeing beyond the complexity of objects and phenomena confronting them. Buffon explained that when 'on jette pour la première fois les yeux sur ce magasin rempli de choses diverses, nouvelles & étrangères, la première sensation qui en résulte, est un étonnement mêlé d'admiration, & la première réflexion qui suit, est un retour humiliant sur nous-mêmes. On ne s'imagine pas qu'on puisse avec le temps parvenir au point de reconnoître tous ces différens objets.'[14] The synchronization of thought and nature was neither immediate nor instinctive; Buffon made it clear that people did not innately know how to see. On first glance, nature resembled not a beautiful, preordained order but a chaos, like a disorganized warehouse or a destroyed museum: randomly scattered pieces, objects and fragments with neither individual distinction nor general order. However, Buffon believed that this first impression did not show nature as it truly was, nor did it embody the nature people would

13. On how Buffon's 'Premier discours' was influenced by and a response to Descartes's *Discours de la méthode*, see Roger, *Buffon*, ch.6.
14. '[O]ne glances for the first time at this storehouse filled with diverse, new and strange things, the first sensation that results is shock punctuated by admiration, and the first reflection that follows is a humiliating retreat back into ourselves. One does not imagine that, with time, it is possible to reach the point of being able to recognize all these different objects.' Buffon, *Histoire naturelle* (1749-1767), vol.1, p.5.

eventually come to know through empirical observations, as by the reading of the *Histoire naturelle.*

Upon first glance at the surrounding world, 'nous ne découvrons en tout cela aucune régularité, aucun ordre' ('we do not find in this any regularity or any order'). There seemed to be only a great chaos of things, 'placées comme au hazard & sans aucune règle apparente; [...] toutes mêlées & dans une espèce de confusion qui ne nous présente d'autre image que celle d'un amas de débris & d'un monde en ruine' ('placed as if by chance and lacking any apparent logic [...] all mixed up and in a sort of confusion that gives us no other image than that of a pile of debris and of a world in ruins').[15] However, the images of a 'pile of debris' and a 'world in ruins' implied that something more constructed had been there in the past, and that some event or process – presently hidden from human perception – had caused its destruction. The ruins attested to what remained from prior events.[16] They were the signature of the temporality of nature, and they made historical investigation, and thus narrative history, possible because they had to be reconstructed – formed into monuments – through observation, thought and language. Ruins were thus the precursors to the 'immortal monuments' which exemplified both natural and stylistic activity. The historian's interpretation of nature would in large part result from his restoration of these ruins, reinserting them back into causal relationships with other events over time. He would breathe the heat of nature back into these fragments and debris, making them whole and living again in the text. It was in this way that the seeming clutter and disarray of the Earth's makeup were, in actual fact, useful and important to the production of the natural historical text. From this initial chaos, the historian was compelled to spend more time making observations, to make his descriptions more attentive and to dedicate himself to building an order which would eventually capture the vital energy of nature. Emphasizing the temporal process of his method, Buffon instructed:

15. Buffon, *Histoire naturelle* (1749-1767), vol.1, p.68-69.
16. English naturalists such as Thomas Burnet believed that the biblical flood was an actual event that had destroyed paradise, leaving 'broken or confus'd heaps of bodies, plac'd in no order to one another', 'the image [...] of a great Ruine, [...] the true aspect of a World lying in its rubbish'. See Burnet, *Telluris theoria sacra: the theory of the earth, containing an account of the original of the earth, and of all the general changes which it hath already undergone, or is to undergo till the consummation of all things* (London, 1691), p.110.

> *Ne nous pressons pas* de prononcer sur l'irrégularité que nous voyons à la surface de la terre, & sur le désordre apparent qui se trouve dans son intérieur, car nous en reconnoîtrons bien-tôt l'utilité & même la nécessité; & *en y faisant plus d'attention* nous y trouverons peut-être un ordre que nous ne soupçonnons pas, & des rapports généraux que nous n'apercevons pas au premier coup d'œil.

> Let us not hurry to judge the irregularity that we see upon the surface of the earth and the apparent disorder that is found in its interior, for we will soon recognize their utility and even their necessity, and by paying more attention we will perhaps find an order that we do not suspect, and general relationships that we do not catch at first glance.[17]

Buffon's method operated upon this principle of the *mise-en-scène* of nature's story through the gradual, temporal process of discovering relationships. This process was evident in his vocabulary in the 'Premier discours': one had to 'se familiaris[er] [avec]' ('familiarize oneself [with]') the surrounding world, '[le] voir *beaucoup* & revoir *souvent*' ('see [it] a great deal and go back to it often'). This formed '*peu à peu* des impressions *durables*' ('little by little lasting impressions') and finally, with the discovery of the most general causes possible, established 'des rapports *fixes & invariables*' ('fixed and invariable relationships'). Moreover, studying nature through time put the observer '*en état* d'étudier avec ordre, de réfléchir avec fruit' ('able to study with order, to reflect fruitfully').[18] It trained the mind to be able to operate like nature, as Buffon perceived it: a focus not on particular events but on the continuous natural process or dynamism which created and ordered the world.

Buffon transformed the initial confusion and shock in reaction to the immensity and heterogeneity of nature into a productive process, 'mêlé d'admiration' ('punctuated by admiration'). Admiration here appeared to be part of a rational method. The admiring mind overcame a preliminary destabilizing experience, gradually moving into dialogue and understanding with that which was once unfamiliar and overpowering. Buffon's use of the idea of admiration was similar to a materialist model developed by Diderot. In the *Encyclopédie* Diderot defined it as the process of learning to understand an outside object or phenomenon by sensing the vibration of one's being on the same

17. Buffon, *Histoire naturelle* (1749-1767), vol.1, p.69-70 (my emphasis).
18. Buffon, *Histoire naturelle* (1749-1767), vol.1, p.6 (my emphasis).

wavelength as that of the object of admiration. In an admiring state the observer did not lose his or her own identity or autonomy but rather refined these with respect to the new aspects of the other object or body. The feeling of admiration was continuous, always unfolding and always providing new ways for the ever-admiring observer to relate to the object. This was especially the case with an elaborate and immense object such as nature: Diderot wrote that 'si l'objet est tel que plus nous l'examinons, plus nous y découvrons de perfections, l'admiration augmente' ('admiration grows if the object is such that the more we examine it, the more we discover its perfections').[19] For those observers and authors who followed this process, nature was the perfect palette upon which to mix, and to ameliorate, the colors of their perception. The more they admired the natural world, Diderot implied, the more beautiful and harmonious their expression of it could become.

Faced with the frightening and seemingly overpowering 'storehouse filled with diverse things' ('magasin rempli de choses diverses') of the world, Buffon expected his readers to slowly learn to temper their astonishment, confusion or fear through admiration. He transmitted the sensation and experience of admiration in the text through lyrical language – much like the verbal painting of the horse described in the previous chapter. In his 'Second discours' (1749), for instance, after again having begun with a depiction of the apparent disorder of the world, he followed it calmly and melodically with this passage:

> Cependant nous habitons ces ruines avec une entière sécurité; les générations d'hommes, d'animaux, de plantes se succèdent sans interruption, la terre fournit abondamment à leur subsistance; la mer a des limites & des loix, ses mouvemens y sont assujétis, l'air a ses courans réglez, les saisons ont leurs retours périodiques & certains, la verdure n'a jamais manqué de succéder aux frimats: tout nous paroît être dans l'ordre; la terre qui tout à l'heure n'étoit qu'un cahos, est un séjour délicieux où règnent le calme & l'harmonie, où tout est animé & conduit avec une puissance & une intelligence qui nous remplissent d'admiration & nous élèvent jusqu'au Créateur.

> Yet we inhabit these ruins with complete assurance. Generations of men, animals, and plants succeed one another without interruption, the earth provides abundantly for their subsistence; the sea has limits and laws to which its movements are subject, the air has its regulated currents, the seasons have their periodic and definite revolutions,

19. Diderot, 'Admiration', in Diderot and D'Alembert, *Encyclopédie*, vol.1, p.140.

> greenery never fails to come after cold weather: everything seems to us to be in order. The earth, which just a while ago was nothing but a chaos, is a charming destination where calm and harmony reign, where all is animated and led with a power and an intelligence that fill us with admiration and that raise us to the level of the Creator.[20]

As in the passages from the article about horses, here the rhythm and flow of language – its heat – indicated the author's purpose: to guide the reader out of the 'ruins' and into a state in which he or she might begin to recognize the order of the world, governed and choreographed in regular movements by the laws of nature. The parallel structure of the first sentence of the above passage, both reassuring and instructive, sought not only to persuade the reader of the reliability and consistent reoccurrence of natural phenomena, but also to explain how to discover this reliability, the hidden and fundamental mechanics of nature, through consistent observation of the world over time. The work of the author who possessed style was thus to combine the scattered pieces of the world into a whole using the method of the 'Discours sur le style'. Reality and its textual representation gradually came to behave in the same way and with the same vitality. Moreover, the harmony that this process of observation created between nature and Buffon's perception of it allowed the reader to see from the perspective of the Creator – who appeared in the above passage to be conflated with the author of the natural historical text. It was from this privileged position, guided by the voice of the author, that readers could learn to visualize the general relationships that Buffon thought constituted nature, and to conceptualize the chains of cause and effect that structured both its laws and its story.

From style to history: reading temporality into nature's story

Through the temporal process of observation, Buffon created a language that reenacted nature's work. Just as events happened over time, so his language unfolded over time as would a story. An author could control when an imagined story started and ended, and how the parts flowed one from another. But what happened when telling the story of nature? Where did it begin and end, and what did it mean to perceive it as a whole?

20. Buffon, *Histoire naturelle* (1749-1767), vol.1, p.69.

The 'Premier discours' argued that, as people made sense of the ruins of nature through observations and the formation of *rapports*, the history of nature would become ever more legible, as well as more detailed and longer, stretching further back in time. In his 'Essais d'arithmétique morale', Buffon explained this process which proceeded, through constant experience and the creation of analogies, from the specific and local to the general and comprehensive:

> [L]orsque nous voyons un phénomène nouveau, un effet dans la Nature encore inconnu, comme nous en ignorons les causes, et qu'elles peuvent être constantes ou variables, permanentes ou intermittentes, naturelles ou accidentelles, nous n'avons d'autres moyens pour acquérir la certitude, que l'expérience réitérée aussi souvent qu'il est nécessaire; ici rien ne dépend de nous, et *nous ne connoissons qu'autant que nous expérimentons*; nous ne sommes assurés que par l'effet même et *par la répétition de l'effet.*
>
> [W]hen we see a new phenomenon, a yet unknown natural effect, as we are ignorant of its causes, which can be constant or variable, permanent or intermittent, natural or accidental, we have no other way to acquire certainty than experience, reiterated as often as necessary; here, nothing depends upon us, and we only know inasmuch as we experiment; we are only assured by the effect itself and by the repetition of the effect.[21]

Knowledge was the result of the repeated observation of regular phenomena. In the 'Premier discours', Buffon envisioned the correspondence between related, repeated events in the world (directed by natural laws) and what he called 'analogies', or the relationships between ideas in the mind (directed by the rules of method, recalling the discussion of Descartes in Chapter 1):

> [I]l faut tâcher de s'élever à quelque chose de plus grand & plus digne encore de nous occuper, c'est de combiner les observations, de généraliser les faits, de les lier ensemble par la force des analogies, & de tâcher d'arriver à ce haut degré de connoissances où nous pouvons juger que les effets particuliers dépendent d'effets plus généraux, où nous pouvons comparer la Nature avec elle-même dans ses grandes opérations.
>
> It is necessary to try to raise ourselves toward something larger and more worthy of our attention, and that is to combine our observations,

21. Buffon, 'Essais d'arithmétique morale', p.62 (my emphasis).

> to generalize facts, to link them together by the force of analogies, and to try to attain that high degree of knowledge from which we can judge that local effects depend upon more general effects, and from which we can compare nature with itself in its grand operations.[22]

Over the course of this process, 'nous nous élevons à des vûes plus générales, par lesquelles nous pouvons embrasser à la fois plusieurs objets différens' ('we raise ourselves to more general views, from which we can comprehend several different objects at the same time').[23] When Buffon's story of nature reached the point where he could see the entire network of relationships, the whole world viewed from above and in each individual object that it contained, and when his language articulated the most general forces of nature, his method of observation would have come into harmony with the activity of natural law. Seeing these two mechanisms, or indeed these two poetics, side by side, he could thus 'compare nature [the outside world] with itself [the internal mind]'.

Furthermore, Buffon's text revealed the empirical development of a practice of interpreting the world. As it followed over time the activities of natural phenomena, geological and biological, Buffon's language simultaneously made visible the history of its creation and the process by which it was created. Buffon's concept of natural history combined two meanings of the word 'history': the events that happened in the world and the story of these events. The former type, which I will call 'event history', may be further divided into singular events and the succession of events, or the rationale by which a series of singular events were related to one another. The latter type, 'narrative history', by contrast, referred to the human perception and discourse of events. It might similarly be divided into the description of singular events, and the large-scale perception and narration of the links between events over a long period.

In Buffon's later 'Epoques de la nature' (1778), the particular location in time in the Earth's history at which events took place (i.e. the occurrence and description of singular events, such as volcanic eruptions, floods and the splitting of continents) was key to knowing how to read both his work and the natural world (see Chapter 5). However, in 1749 he did not structure natural history in terms of these kinds of singular events. Changes in the Earth's terrain, for

22. Buffon, *Histoire naturelle* (1749-1767), vol.1, p.51.
23. Buffon, *Histoire naturelle* (1749-1767), vol.1, p.5.

instance, were not identified by the particular moment at which they took place or by their necessary effect on and relationship to previous and subsequent events. Rather, they were understood according to the dual concept of history as the succession of events and the narrative of this succession. The focus was on the scale of the *longue durée*: on the processes through which the Earth was modified over time, which gradually gave it its present-day state and continued to allow its evolution. Buffon looked for a new way to define the relationships between natural phenomena such that they would be commensurable with the relationship between ideas in the mind, making nature comprehensible and legible to human reason. The perception of natural events in the *Histoire naturelle* became a discursive simulacrum of the history of the succession of these events in the world.

At the beginning of the 'Second discours', Buffon stated that 'un Historien est fait pour décrire & non pour inventer' ('a historian is trained to describe, and not to invent'): he can 'examine' and 'represent' the surface of the Earth, but 'il ne doit se permettre aucune supposition' ('he must not allow himself to make any suppositions').[24] This statement, recalling the definition of history discussed in previous chapters, initially appears to contradict the poetic, creative manner in which Buffon discussed style and the re-presentation of nature through the world-building qualities of his language. Yet description in the *Histoire naturelle* was not simply the compilation of a catalogue of all phenomena and objects observed. It was moreover the act of giving order and vitality to these things. Within the act of ordering, the historian might indeed 'faire usage de son imagination' ('make use of his imagination'). He could picture the natural objects in his mind in order to 'combiner les observations, généraliser les faits, & en former un ensemble qui présente à l'esprit un ordre méthodique d'idées claires & de rapports suivis & vrai-semblables' ('combine observations, generalize facts and form from this a whole which presents the mind with a methodical order of clear ideas and consecutive, verisimilar relationships').[25] When telling the geological story of the Earth in the 'Second discours', Buffon first presented the history, or description, of the Earth's physical attributes (oceans, mountains and continents) and then what he called the 'theory', or the attempt to find a physical explanation for these elements.[26] He concluded that to see nature

24. Buffon, *Histoire naturelle* (1749-1767), vol.1, p.68-70.

25. Buffon, *Histoire naturelle* (1749-1767), vol.1, p.68.

26. In the 'Second discours', theory embodies 'une tentative d'explication de ces

from above meant to read it through the movement of the force of water across the surface of the Earth over time. This rhythmic, cyclical process had written the planet's history. Its constant 'flux' and 'reflux' had built continents, mountains and valleys in some periods of history, and had eroded them away in others.[27] This recurrent process continued into the present day, and would continue into the future. Buffon theorized that the ocean's waters would eventually wear away enough of the present continents so as to 'rendr[e] un jour cette terre à la mer, qui s'en emparera successivement, en laissant à découvert de nouveaux continens entre-coupez de vallons & de montagnes, & tout semblables à ceux que nous habitons aujourd'hui' ('one day give this earth back to the sea, which will gradually take possession of it, uncovering new continents broken by valleys and mountains, and all resembling those which we inhabit today').[28]

The historical nature of Buffon's 1749 work has been the subject of scholarly debate, precisely because the history of singular, identifiable and unalterable events in the past did not appear in his early studies of the Earth. Because of this absence, some scholars have argued that temporality is not figured in the text. Jacques Roger claims that Buffon's early work was *not* historical because it explained the present in the same manner, and through the same physical processes and laws, as the past. Any event could have happened at any point in time in Buffon's world, according to Roger, because, as the story was circular, it could not and did not produce difference and historical change – or, if there were change, it always appeared to be subsumed back into sameness. For Roger, history is defined as the 'succession d'événements irréversibles' ('sequence of irreversible events'), resulting in the current state of the Earth as unique in its place and time

caractéristiques par des causes physiques' ('an attempt to explain these characteristics through physical causes'). History and theory are not two distinct ways of knowing, but rather stages of the same process. Buffon 'ne fait donc que rappeler ici les deux moments de la connaissance: la collection des faits et l'élévation aux vues générales' ('thus simply reminds us here of the two moments of knowledge: the collection of facts and the elevation to general views'). See Buffon, *Histoire naturelle*, ed. Schmitt and Crémière, 'Second discours: histoire et théorie de la terre', vol.1, p.229-300 (p.229, n.4).

27. Martin Rudwick characterizes Buffon's 1749 story of geological nature as a 'continuous flux' or a 'dynamic equilibrium', created by the movement of water, in which 'any particular point on the globe might have been both land and sea at different times.' See Rudwick, *Bursting the limits of time: the reconstruction of geohistory in the age of revolution* (Chicago, IL, 2005), p.140.
28. Buffon, *Histoire naturelle* (1749-1767), vol.1, p.124.

and as defined by all the causally related events that preceded it. He argues that, since the geological story of the Earth in Buffon's 'Second discours' did not produce permanent and legible historical change, it could not be understood in terms of this definition.[29] From Roger's point of view, Buffon's 'Second discours' 'ne contient aucune allusion à une histoire possible de la Nature, ni même au rôle du temps dans les processus naturels' ('contains no allusion to a possible history of Nature, nor to the role of time in natural processes').[30] John Eddy takes an even stronger stand on the issue, claiming that the entire corpus of Buffon's work was 'fundamentally nonhistorical' and 'nondevelopmental' because it represented the world as eternal, based on the unchanging and fixed character of Newtonian natural law. Eddy argues that '[e]very part of the *Histoire* reveals Buffon's enduring conviction that within the deep recesses of nature, […] where the forces and mechanisms of nature operate, the Heraclitean flux of sensible phenomena hides an almost Parmenidean eternity.'[31]

However, Roger's and Eddy's arguments apply a modern conception of history to the *Histoire naturelle.* Buffon's work was not any less historical because his world seemed not to change, or because his narrative was not fundamentally based on a contingent, causal progression of events. From this point of view, Jean Goulemot's work on the perception of history in the seventeenth and eighteenth centuries helps to modify the argument that the *Histoire naturelle* was ahistorical. In histories from this time, according to Goulemot, the appearance of cycles or revolutions allowed the reader to 'trouver les lois du devenir historique, et […] [de] saisir, au-delà du particulier, le général' ('to find the laws of historical development, and […] to grasp the general from beyond the particular').[32] Perceiving the significance of each individual event was part of a larger activity of understanding the forces which allowed for history to take place, forces that focused upon 'l'existence et non le contenu de l'historiographie des révolutions' ('the existence and not the content of the historiography

29. Jacques Roger, 'Buffon et l'introduction de l'histoire dans l'*Histoire naturelle*', in *Buffon 88*, ed. J.-C. Beaune *et al.*, p.193-205 (194).
30. Roger, 'Buffon et l'introduction de l'histoire', p.197.
31. John Eddy, 'Buffon's *Histoire naturelle*: history? A critique of recent interpretations', *Isis* 85:4 (December 1994), p.644-61 (645). On the idea of circular history in the Enlightenment, see Nathaniel Wolloch, *History and nature in the Enlightenment: praise of the mastery of nature in eighteenth-century historical literature* (Farnham, 2011) p.210-17.
32. Goulemot, *Le Règne de l'histoire*, p.141.

of revolutions').[33] 'L'important n'y est pas le factuel, la vérité du détail, mais le sens à donner à un vaste ensemble' ('What is important is not the factual, the truth of details, but the meaning to be given to a larger whole'), the story seen as a whole entity.[34] Buffon's history of the Earth might not have had a clear end – in fact, it appeared to be endless, and this was indeed quite a controversial move for his time – but this did not make it less of a history, and thus any less amenable to a historical, temporal narrative.

Buffon began his 'Second discours' with the statement: 'L'histoire générale de la terre doit précéder l'histoire particulière de ses productions' ('The general history of the earth must precede the specific history of its productions').[35] This signifies that, in fact, time did matter to Buffon's sense of history. The generality of which he spoke here was a temporal one: it encompassed all the events that had occurred in nature and it was structured by the causal relationships between these events. According to Buffon, finding the order of the world's past was not only about observing 'faits singuliers' ('singular facts') and local phenomena, but about detecting the 'résultats généraux des observations qu'on a faites sur les différentes matières qui composent le globe terrestre' ('general results of observations one made of the different forms of matter that make up the terrestrial globe'). Buffon's historical narrative in the 'Second discours' involved seeing 'la Nature en grand' ('Nature on a grand scale') and seeking to understand how this scale implied the temporality of a *longue durée*.[36]

Furthermore, to speak of long-term temporality included the idea of a forward, progressive movement that was factored into the circular repetitions of the flow of water. To an older idea of revolutions as periods which circled back to their starting point (like the movement of a planet in its orbit), Buffon added the idea of a gradual development or process of change into the future, producing a dimensional narrative that could be said to take the shape of a spiral. The idea of moving away from the starting point at the end of every revolutionary cycle was introduced. This distancing effect allowed Buffon and his contemporaries to speak of monuments, the residues of previous ages which revealed and made comprehensible the idea of difference from, not sameness with, the past. His story allowed the present-day observer to look back into time and

33. Goulemot, *Le Règne de l'histoire*, p.141.
34. Goulemot, *Le Règne de l'histoire*, p.158.
35. Buffon, *Histoire naturelle* (1749-1767), vol.1, p.65.
36. Buffon, *Histoire naturelle* (1749-1767), vol.1, p.66.

to construct a narrative which, without a sense of causal temporality, would be flat and circular. Therefore, although the 'Second discours' told the Earth's story in terms of repeated, predictable cycles (of rising and receding water), it was not atemporal.[37] Buffon's method of consistent observation relied upon the passage of time in order for the natural historian to construct relationships between objects in the natural world, and thus between ideas in the mind. Knowledge, too, was an event in the world produced through time. As a result, in the process of writing a historical narrative, the causal events involved in constructing a more general argument appeared to coincide with the material events in the world which were linked to describe natural laws. The writing of narrative history in the *Histoire naturelle* was the result of the combination of the continuity of natural events through time with the temporal process of observing and reasoning about these events – and the two were tied together by the written language of the author.

This conflation of the two meanings of history, real events and written story, was an issue of which I believe Buffon was quite aware. In a text very different from the 'Premier' and 'Second discours', the 'Essais d'arithmétique morale', Buffon confronted the mathematics of belief, trying to determine at what point human certainty of an event in the world could be spoken of as truth. He emphasized the intersection of forces, the asymptotic line along which the activity of the mind became more and more like the activity of nature:

> Dès qu'[un événement naturel] sera arrivé treize ou quatorze fois de la même façon, nous avons déjà un degré de probabilité égal à la certitude morale qu'il arrivera de même une quinzième fois, et de ce point nous pouvons bientôt *franchir un intervalle immense*, et conclure par analogie que cet effet dépend des loix générales de la Nature, qu'il est *par conséquent aussi ancien* que tous les autres effets, et qu'il y a certitude physique qu'il arrivera toujours comme il est toujours arrivé, et qu'il ne lui manquoit que d'avoir été observé.

37. Another aspect of the debate over the historical nature of Buffon's work is raised by Thierry Hoquet, who maintains that eighteenth-century history writing remained 'a method of description [...], and the question of time had little or nothing to do with it'. 'Description', the 'more or less nonsystematic collection of data', was paired with the practice of civil historians who 'acquir[ed] certainty through the search for specific evidence', and decided (especially through 'the authority of testimony') which facts carried the most assurance. See Thierry Hoquet, 'History without time: Buffon's *Natural history* as a nonmathematical physique', *Isis* 101 (2010), p.30-61 (44-45).

> As soon as [a natural event] has happened thirteen or fourteen times in the same way, we have already a degree of probability equal to the moral certainty that it will happen the same way a fifteenth time, and from here we can soon *cross an immense interval,* and conclude by analogy that this effect depends upon the general laws of Nature, that it is *consequently as old* as all the other effects, and that there is a physical certainty that it will always happen as it has always happened, and that it only needed to have been observed.[38]

This passage reveals Buffon's understanding of the relationship of observation to the production of a clear body of knowledge, as well as to the construction of a coherent narrative history of what was being observed. When observations of a particular event acquired so much probability that they became 'moral certainties' (that is, considered to be true) the mind then was enabled to make a conceptual and intellectual leap to understanding this event as an effect of a general law of nature. This event, of which people were certain, and which they had rendered causal by logically linking it to a natural law, was 'consequently as old as all the other effects', and was renamed a 'physical certainty', a truth of nature. In the word 'old' ('ancien'), Buffon conflated logical and natural processes. The resulting science was dependent not only upon the succession of the historian's observations of the world, but also upon another dimension, historical depth – the forward-moving vector of the spiral conception of history. The construction of a chain of reasoning and argumentation was simultaneously, and according to the same rules, the construction of a chain of events that led Buffon's thought back in time into nature's past. The process of discovering and inducing natural law, of finding the summit of a chain of observations and the point which embodied all figures of relationship and analogy below it, was akin to moving back in time and identifying the laws which had always given regularity and consistency to physical phenomena. The highest, most general, point of nature's structure was also its oldest. From it were derived both the natural predictability and the human certainty that an event 'will happen as it has always happened'.

Even if it had no clear end or finality, Buffon's story had a beginning (the story of the comet, addressed in the following chapter). In order to tell a story, he did not have to tell the whole story. As long as the greater structuring principles or laws of the narrative

38. Buffon, 'Essais d'arithmétique morale', p.63 (my emphasis).

were made clear, it was still a history, even if it remained unfinished because the causal series of events remained open-ended. The spiral nature of historical causality, moreover, did not imply illegibility. Rather, it enacted a different set of rules for perception, and in turn a different way of reading and interpreting law. Enlightenment history was the pursuit of a narrative which accounted for the past in order to structure the future. Jean Goulemot names this process the 'mise en situation' ('situating') of an event or object 'dans un réseau de causes et de conséquences' ('in a network of causes and consequences'), and the devising of a discourse that grants 'lisibilité' ('legibility') to this event by emphasizing the process or narrative of which it was a part.[39] The rules of style made Buffon's discourse of the world legible in a manner very different from Newton's, for instance. The 'Premier discours' showed, in the words of Goulemot, 'le procès d'édification de l'histoire qui constitue l'événement ou la série en un texte lisible' ('the process of historical edification which forms the event or the sequence [of events] into a legible text').[40] History was not simply or necessarily the unearthing, or the 'dévoilement', of forgotten past events or memories. It was the assembly of a narrative based upon method and logical structures of analogy. For Goulemot, '[L'histoire] est construction, à partir d'un savoir idéologique qui la précède, qu'elle sert à fonder et sans lequel elle ne pourrait s'énoncer. L'histoire n'est pas hors des discours qui la constituent [...]. Elle est procès d'édification' ('[History] is construction, starting from a preceding ideological knowledge, that it founds and without which it could not formulate itself. History is not outside of the discourses that constitute it [...]. It is a process of edification').[41]

This legibility in Buffon's narrative of the duration of events is evident in the 'Second discours'. The experience of reading his story was designed to make the reader feel as if he or she were floating over the surface of the Earth, digging into its core or swimming under the

39. Goulemot, *Le Règne de l'histoire*, p.25-26.
40. Goulemot, *Le Règne de l'histoire*, p.28.
41. Goulemot, *Le Règne de l'histoire*, p.25-26. In Jordanova's words, 'Perhaps the most important change in the idea of nature during [...] the eighteenth and nineteenth centuries [...] is the articulation and consideration of the idea that nature itself is inherently historical [...]. The sense of the historicity of nature not only called into questions assumptions about stability and permanence, but also raised some tricky epistemological questions about how nature's past could be known about [...]. [A] whole host of imaginative possibilities were opened up by the discovery of time as an agent in nature' ('Introduction', p.28).

sea. The reader's thoughts were produced guided by Buffon's voice, asking him or her to form analogies with previous descriptions, and to observe and to feel how one kind of landscape seemed to change and flow smoothly into another:

> Considérant ensuite le fond de la mer, nous y remarquons autant d'inégalités que sur la surface de la terre; nous y trouvons des hauteurs, des vallées, des plaines, des profondeurs, des rochers, des terreins de toute espèce; nous voyons que toutes les isles ne sont que les sommets de vastes montagnes [...]. Là sont ces contrées orageuses où les vents en fureur précipitent la tempête, où la mer & le ciel également agitez se choquent & se confondent: ici sont des mouvemens intestins, des bouillonnemens, des trombes & des agitations extraordinaires causées par des volcans dont la bouche submergée vomit le feu du sein des ondes [...]. Plus loin je vois ces gouffres dont on n'ose approcher, qui semblent attirer les vaisseaux pour les engloutir: au delà j'aperçois ces vastes plaines toûjours calmes & tranquilles [...]; enfin portant les yeux jusqu'aux extrémités du globe, je vois ces glaces énormes qui se détachent des continens des poles, et viennent comme des montagnes flottantes voyager & se fondre jusque dans les régions témpérées.

> Considering next the bottom of the sea, we notice as many irregularities there as on the surface of the earth; we find peaks, valleys, plains, depths, boulders, all sorts of terrains; we see that islands are nothing but the summits of vast mountains [...]. There are those tempestuous regions where violent winds hasten storms, where the sea and the sky, equally agitated, collide and become confused with one another: here are the internal movements, turbulences, torrents and extraordinary unrest caused by volcanoes whose submerged mouths vomit fire from the depths of the waves [...]. Further along I see those abysses which one does not dare approach, which seem to attract and swallow ships; beyond I glimpse those vast plains, always calm and tranquil [...]; finally, casting my gaze toward the extremities of the globe, I see those enormous glaciers that separate from the continents of the poles and travel, like floating mountains, all the way to the temperate regions to melt.[42]

Here is a passage that operates simultaneously in imaginative, physical and temporal space. Perhaps Buffon did not speak of specific events in the Earth's past, with precise dates, but as his readers voyaged to the different geological realms of the planet, they could sense the general movement of the broad forces of nature as these forces currently

42. Buffon, *Histoire naturelle* (1749-1767), vol.1, p.71-72.

operated and as they had worked for many years: the movement of the sea and of submerged continents, the work of fire and ice to contribute to the formation and changes of both the land and the sea. This passage recalls the analysis of heat as a new way of not only perceiving what currently existed, but grasping the power of nature through Buffon's language in order to visualize and to narrate what had happened and what would happen. The early volumes of the *Histoire naturelle* were indeed historical, because they told stories that constantly searched for more and more general explanations within deep temporality. The 'Second discours' was a *mise-en-scène* of this process.

Reading the *Histoire naturelle* was a practice of learning how to see differently, and specifically to see with historical eyes – something Buffon's readers may not have known innately how to do. To the traditional sense of history as description was added an element of expansion and dynamism, which could only be sensed and understood by becoming literate in reading the world as well as the narrative of it, both of which were successive and temporal processes. Style allowed these stories to be told in a manner which changed what it meant to see and to experience time, a manner which probed nature's inner mechanisms, transmitting the operation of laws through the pulse of language.

4. Hypothesis and the energy of invention

Only the artist who constantly brings forth from within himself worlds in miniature giving them definite shape will be able to understand the universe as the creation of the same forces of which he is aware in his own creative process.[1]

[A]rrestez-vous encore un peu à considérer ce cahos, & remarquez qu'il ne contient aucune chose qui ne vous soit si parfaitement connuë, que vous ne sçauriez pas mesme feindre de l'ignorer. Car pour les qualitez que j'y ai mises, [...] je les ay seulement supposées, telles que vous les pouviez imaginer [...]. Et son idée [de la matière] est tellement comprise en toutes celles que nôtre imagination peut former, qu'il faut nécessairement que vous la conceviez, ou que vous n'imaginiez jamais aucune chose.

[L]inger a bit longer to consider this chaos and to notice that it contains nothing which you do not know perfectly, that you would not even be able to pretend to ignore it. For, in terms of the qualities that I placed there, if you paid attention, I only imagined them as you would imagine them [...]. And the idea [of matter in this chaos] is so much a part of all those [ideas] that our imagination can make, that it is

1. Cassirer, *The Philosophy of the Enlightenment*, p.317.

> necessary that you conceive of it, or you would never imagine anything.[2]

In the opening essays of the first volume of the *Histoire naturelle*, Buffon wrote of natural history using a temporal narrative of recurrent, cyclical processes. He used observations of the present state of the Earth to theorize that, over many centuries, the Earth endured and was shaped by the periodic movement of water and the constant processes of erosion and sedimentation, which eternally built up and wore away at continents.

The next part of this first volume was striking, as it appeared incongruous with the constant, predictive, repetitive theory of the flow of water. In this essay, entitled 'Preuves de la théorie de la terre. De la formation des planètes' ('Proofs of the theory of the Earth. On the formation of the planets'), Buffon imagined traveling far back in geological time to conjecture upon that which was not directly observable to the eye: the origin of the planet Earth within the origin of the planetary system as a whole. Buffon tried to make this event otherwise visible, through the image formed by the heat of his words in readers' minds. He painted a story of the formation of the planets, originating from the matter ejected from the impact of a comet into the Sun. After the very gradual description and rationalization of the current form and state of the Earth, created by the enduring and repetitive ebb and flow of water, Buffon suddenly confronted his readers with the immediacy of a comet smashing into the Sun and ripping from it a part of its mass, and with the violence and irreversibility of a world shaped by heat and fire.

Instead of contradicting the process of bringing the natural historical text into harmony with the behavior of nature, the story of the comet did in fact tell not only of the origins of the world but also of those of knowledge and language. Buffon showed how from a preliminary and singular event, a balance was restored and maintained between the events of history and the narrative of the natural historian. The enduring energy, the immortal monument of style discussed in the previous chapter, was produced from the initial possibility of the hypothetical. This hypothesis took on a new, hybrid form – not the same as the use of the term today, nor quite resembling

2. René Descartes, *Le Monde de M. Descartes, ou le Traité de la lumière et des autres principaux objets des sens* (Paris, Chez Michel Bobin et Nicolas Le Gras, 1664), p.74-75.

the Cartesian or Newtonian uses that already existed in the 1750s. Buffon employed hypothesis not as part of what might be called the scientific method, but as the foundation of a discourse which sought to persuade the reader of the validity and rigor of his historical *and* stylistic system.

Buffon began the 'Preuves de la théorie de la terre' by distinguishing what he accomplished in the previous chapters from the kind of work he would now be doing. This distinction, he hoped, would

> mettre le lecteur plus en état de prononcer sur la grande différence qu'il y a entre une hypothèse où il n'entre que des possibilités, & une théorie fondée sur des faits, entre un système tel que nous allons en donner un dans cet article sur la formation & le premier état de la terre, & une histoire physique de son état actuel, telle que nous venons de la donner dans le discours précédent.

> put the reader in a better state to judge the significant difference between a hypothesis, where there are only possibilities, and a theory founded on facts; between a system such as we will present in this article on the formation and the original state of the Earth, and a physical history of its actual state such as we just presented in the preceding discourse.[3]

Theory, 'founded on facts', was related to the work of history accomplished in the preceding chapters of the *Histoire naturelle*. The collection and depiction of repeated observations allowed for the formation of facts, information considered to be part of natural knowledge because based upon a large quantity of similar observations and analogies amassed over time. Facts were assembled into a theory seeking to identify their larger causes and laws. By contrast, Buffon argued that a hypothesis was founded upon 'possibilities' and that it was used to build a 'system'.[4] In one contemporary definition from the *Encyclopédie*, a system was described as 'la disposition des différentes parties d'un art ou d'une science dans un état où elles se soutiennent

3. Buffon, *Histoire naturelle* (1749-1767), vol.1, p.129.
4. 'System' was an ambiguous and complicated term at the time, employed in authors' critiques of one another's work. These critiques usually centered on the accusation that the work was not clearly based in empirical observations and thus that it was too 'Cartesian'. According to Jeff Loveland, systems were also associated with 'teaching, schools, pedantry, and scholasticism', and 'were seen as dogmatic and irrelevant intellectual whimsies'. See Loveland, *Rhetoric and natural history*, p.101-108.

toutes mutuellement, & où les dernières s'expliquent par les premières' ('the disposition of different parts of an art or a science into a state in which they all mutually support each other, & in which the final parts are explainable via the initial ones').[5] Interestingly, the truth value of a system was not emphasized in this definition; importance was placed instead on the linking and interdependency of its parts.[6] A system could be ordered and governed by mental and logical rules which gave measurement and proportion to all of its parts, putting them in relationship to one another – as described in a work such as Descartes's *Règles pour la direction de l'esprit.* Understanding a system in this way was less about determining its specific truth value than about explaining how it worked, by comprehending how the individual parts were proportioned and ordered to create a whole.

It is thus significant in the above citation that Buffon grouped together the terms 'hypothesis' and 'system'. What he referred to as his hypothesis was an attempt to continue to prove and render persuasive his theory of the Earth, not by claiming to find the truth of the planet's origin, but by revealing the method and rules which had to be followed in order to construct a story in which all elements existed in a proportional relationship to one another and thus appeared verisimilar. These elements were moreover structured according to the idea presented in the 'Discours sur le style' that the structure and operation of text should be in harmony with the events of the world itself. Buffon's purpose seemed to be to persuade his readership, by a rigorous relationship between the text and the world, to accept his hypothesis as valid, and then to use the ensuing conviction and interest to continue the investigation of nature. Thus, what is crucial to understand about the use of hypothesis in the *Histoire naturelle* is neither its truth status nor its ability to be proven through visual observation of the surrounding world, but the language through which it built persuasive worlds. Its strength and its plausibility were a result of energetic discourse which, by using nature as a model, gained the ability to produce like nature, to act as nature would act. This was

5. 'Système', in Diderot and D'Alembert, *Encyclopédie*, vol.15, p.777.
6. The article 'Système' described three types of systems, each one based upon a different kind of evidence. There were those based upon 'des maximes générales ou abstraites' (and thus logically demonstrable). Then, there were those grounded in experience and established facts. Finally, there were those guided by 'suppositions'. However, even though each of these systems was composed of elements of a different epistemological value, this did not change the way in which they were understood to operate. 'Système', p.777.

true even in cases such as the beginning of the universe when it was not possible to directly observe the events. In the end, readers had to judge ('prononcer sur') the inventive power of the hypothetical world.

Hypothesis and the invention of a verisimilar world

That making hypotheses could be understood as a discursive activity in the eighteenth century is suggested in the article 'Hypothèse' from the *Encyclopédie*, in which the term was defined as 'la supposition que l'on fait de certaines choses pour rendre raison de ce que l'on observe, quoique l'on ne soit pas en état de démontrer la vérité de ces suppositions' ('the supposition that one makes about certain things to give reason to what one observes, even though one is not able to demonstrate the truth of these suppositions').[7] A hypothesis did not indicate whether observations were true or false, but instead proportioned them with respect to one another, making them rationally legible to the intellect, and allowing them to be systematically ordered and explained.

The original definition of 'hypothesis', as a term of rhetoric and oratory from Greek, was the founding supposition of an argument, that which lay underneath, 'hypo'.[8] According to the *Encyclopédie* article, hypotheses were 'invented': they contributed to knowledge through 'une méthode très-nécessaire à l'art d'inventer' ('a method very important to the art of invention').[9] The art of invention was linked to the idea of discovery which, as evidenced in the following definitions, included the acts of both finding new things in the world and finding – imagining – new ideas in the mind.[10] In Furetière's *Dictionaire universel* [*sic*], invention was defined as a 'subtilité d'esprit, [un] certain genie

7. 'Hypothèse', in Diderot and D'Alembert, *Encyclopédie*, vol.8, p.417.
8. Furetière defined 'hypothèse': 'Ce mot vient du Grec *hypothesis*, d'*hypo*, *sub*, & de *thesis*, *posito*, *hypothese*, *supposition*' ('This word comes from the Greek *hypothesis*, from *hypo*, *sub*, and from *thesis*, *posit*, *hypothesis*, *supposition*'). It was a 'supposition qu'on fait d'un principe, d'une proposition, pour en tirer des raisonnements & des consequences' ('Supposition that one makes from a principle, from a proposition, in order to draw arguments and consequences'). See Antoine Furetière, *Dictionaire universel* (The Hague, A. et R. Leers, 1690), http://gallica.bnf.fr/ark:/12148/bpt6k50614b (last accessed March 22, 2018), vol.2, p.292.
9. 'Hypothèse', p.418.
10. In his analysis of Buffon's cosmographical thought, Benoît de Baere links invention to an authority on rhetoric in conventional use at the time, Cicero's *De inventione*. See Benoît de Baere, *La Pensée cosmogonique de Buffon: percer la nuit des temps* (Paris, 2004), p.109-10. On Buffon's use of sources from Antiquity,

particulier qui donne la facilité de trouver quelque chose de nouveau' ('subtlety of mind, [a] certain, particular spirit which makes it easy to find something new'); in the *Dictionnaire de l'Académie française*, 'to invent' meant '[t]rouver quelque chose de nouveau par la force de son esprit, de son imagination. *Inventer un Art, une Science. Inventer un systême, une machine*' ('[t]o find something new through the force of one's mind, of one's imagination. *To invent an Art, a Science. To invent a system, a machine*').[11] These definitions were echoed in the first stage of Buffon's 'Discours sur le style': the formulation of a primary plan which contained the seeds of the argument's energy and activity. The art of invention was not that of creation (recall that, in the 'Discours sur le style', Buffon wrote that the human mind could not create), but instead referred to the process of building a discursive simulacrum of the subject matter from an original 'plan', with strong analogies to existing natural phenomena. The hypothesis that supported Buffon's argument about the origins of the Earth was meant to put readers in a state of believing its underlying premise to be true, and then of being receptive to the heat of Buffon's language as he built this premise into a plausible narrative.[12]

The most famous, and most frequently referenced, example of hypothesis in the Enlightenment was René Descartes's *Le Monde, ou le Traité de la lumière* (1664). Descartes imagined the creation of a new world using the laws of his corpuscular theory of matter and its motion. He directly appealed to his readers, guiding them through the stages of development of this world:

> Permettez donc pour un peu de temps à vôtre pensée de sortir hors de ce Monde, pour en venir voir un autre tout nouveau que je feray naître en sa présence, dans les espaces imaginaires [...]. Entrons-y seulement si avant que nous puissions perdre de veuë toutes les créatures, que Dieu fît il y a cinq ou six mille ans, & après nous être arrétez là en quelque lieu déterminé, supposons que Dieu crée de nouveau tout autour de nous tant de matière, que de quelque côté que nôtre imagination se puisse étendre, elle n'y aperçoive plus aucun lieu qui soit vuide.

---

see Buffon, *Histoire naturelle*, ed. Schmitt and Crémière, 'Introduction', vol.8, p.13-131.

11. Furetière, *Dictionaire universel*, vol.2, p.374; 'Invention', in *Dictionnaire de l'Académie française*, n.p.
12. For the enumeration of the steps involved in formulating an argument, see Cicero, *On the ideal orator (De oratore)*, ed. and translated by James M. May and Jakob Wisse (Oxford, 2001), p.153-58.

> Allow, therefore, for a little while, your mind to leave this world, to come to see another completely new one that I will make appear in its presence, in imaginary space [...]. Let us only enter [this world] if, before losing sight of all the creatures that God made five or six thousand years ago, and after stopping in some determined place, we suppose that God creates all over again, all around us, so much matter, that to whatever edge our imagination stretches, it cannot find a single spot that is empty.[13]

In this text, Descartes emphasized the equivalence between the work of God's laws and the work of the human intellect: what happened or could happen in the world was tantamount to that which people were able to imagine. His goal in *Le Monde*, Descartes explained, was not

> d'expliquer comme eux [les Philosophes] les choses qui sont en effet dans le vray monde; mais seulement d'en feindre un à plaisir, dans lequel il n'y ait rien, que le plus grossier esprit ne soit capable de concevoir, & qui puisse toutefois estre créé tout de mesme que je l'auray feint [...]. [I]l est indubitable qu'encore qu'il n'y ait rien de tel dans l'ancien monde, Dieu le peut toutesfois créer dans un nouveau. Car il est certain qu'il peut créer toutes les choses, que nous pouvons imaginer.

> to explain, as they [the Scholastic philosophers] do, the things that are actually in the true world; but only to spontaneously feign one in which there is nothing of which the crudest mind is not capable of conceiving, and which can notwithstanding be created just as I will have feigned it [...]. It is indubitable that, while there be still nothing like it [what Descartes imagined] in the old world, God is nevertheless capable of creating it in a new one. For it is certain that he can create all the things that we are capable of imagining.[14]

Descartes set up an equivalence between the outside world and the intellect such that the physical world was understandable as the product of human invention. As Catherine Labio has argued, because of the theological constraints stipulating that the real world could be no more than 6000 years old, Descartes had to make a dangerous and 'radical fictional leap' in 'an attempt to get at the truth by setting the stories of Scholastic philosophers aside and providing an account that could be mechanistically replicated, even in the absence of God'.[15] The

13. Descartes, *Le Monde*, p.66-68.
14. Descartes, *Le Monde*, p.76-78.
15. Catherine Labio, *Origins and the Enlightenment: aesthetic epistemology from Descartes to Kant* (Ithaca, NY, 2004) p.24.

actual existence of Descartes's new world diminished in importance in relation to the mental process of forming a simulacrum of this world. Validity came from using method to produce chains of reason, linking ideas in the same manner as causes and effects would be linked in the world. The method Descartes used to order his ideas about the invention of a new world in the text thus enabled him to speak about the laws of this world: 'la raison qui me les a enseignées, me semble si forte, que je ne laisserois pas de penser d'estre obligé de les supposer dans le nouveau [monde]' ('reason, which taught them to me, seems to me so strong, that I would not hesitate to think myself obliged to assume them in the new [world]').[16] His argument proceeded not from observation of the real world to the invention of a rational one in the mind, but rather from the fable outwards.[17] Labio writes that 'the Cartesian fable exposes truth by creating it.'[18] In *Le Monde* the laws of nature did not structure the nature and order of human thought. Rather, the ability to speak of and order the laws of nature flowed from the rules of reason.

In the eighteenth century, the Cartesian activity of world-making was recast as an important and still germane example of hypothesis in the *Dictionnaire de l'Académie française*: 'HYPOTHÈSE se dit aussi De l'assemblage de plusieurs choses qu'on imagine & qu'on suppose pour parvenir plus facilement à l'explication de certains Phénomènes; ce qu'on appelle autrement & plus communément, *Système. L'hypothèse de Ptolomée. L'hypothèse de Copernic. L'hypothèse de Descartes.*'[19] The formation of a hypothesis, in which 'we imagine' and 'we suppose' so as to 'more easily' find the answer, no longer appeared to be necessarily associated with the production of useful knowledge about nature, however. By the end of the 1740s, the terms 'system' and 'hypothesis' had become

16. Descartes, *Le Monde*, p.92.
17. Jean-Pierre Cavaillé writes of *Le Monde*: 'l'enjeu de cet invraisemblable saut hors du monde est de faire "paraître" la vérité' ('the stakes of this improbable leap away from the world are to make truth "appear"'). The fable becomes 'une "invention" au sens moderne; c'est-à-dire comme instauration, non restauration de la science' ('an "invention" in the modern sense: that is, as the creation, and not the restoration, of knowledge'). Cavaillé, *Descartes: la fable du monde* (Paris, 1991), p.183-85.
18. Labio, *Origins and the Enlightenment*, p.23.
19. 'HYPOTHESIS is also known as The assemblage of several things that we imagine & that we suppose to more easily arrive at the explanation of certain Phenomena; what we call otherwise, and more familiarly, *System. The hypothesis of Ptolemy. The hypothesis of Copernicus. The hypothesis of Descartes.*' 'Hypothèse', in *Dictionnaire de l'Académie française*, n.p.

expressions of critique and attack.[20] (Buffon's use of the term hypothesis indeed provoked criticism that it was a system or fable with no links to or proofs in reality.)[21] As Descartes's ideas were read through the lenses of authors such as Locke and Newton, it became more difficult to understand how a system so isolated and closed in upon itself, as was the one in *Le Monde*, could be related to and informative of reality. In *Le Monde* this world had its own, perfectly rigorous internal rules, but they were deemed to lack relationship to the study of the events and phenomena of the outside, phenomenal world.

Despite the critiques he received, I believe that Buffon's story of the comet was constructed slightly differently from Descartes's model and thus in some ways resembles a more modern idea of hypothesis, based upon verifiability and proposing an explanation that forms the starting point for further investigation and experiment. Buffon used the elements of natural historical language – measurement, heat, style – to bring the methodological and rational tools of Descartes's internally coherent world into proportion with the search for knowledge of the outside, physical world. The structure of thought could thus be harmonized with the order of nature. Moreover, through style, Buffon pushed this hybrid epistemology one step further, questioning the meanings and uses of certain types of proof in a hypothetical argument. In the search for the general mechanisms and order of nature, Buffon wrote in the 'Essais d'arithmétique morale' that a natural historian's representations of nature would eventually leave the realm of visible, observable phenomena as of certain knowledge, entering that of '[des] probabilités de différens degrés' ('probabilities of different degrees').[22] In this essay, he described how to make tentative steps toward discovering more general causes and explanations, even if they did not yet fit into the current worldview, for indeed 'les causes nous sont & nous seront perpétuellement inconnues' ('causes are and always will be perpetually unknown to us'). Buffon explained causes simply as more comprehensive effects, a perspective that focused on the process of finding *rapports* between 'des effets plus généraux' ('more general effects'), rather than on only looking for the final

20. On Enlightenment debates about these terms see Loveland, *Rhetoric and natural history*, p.101-103.
21. See Buffon, *Histoire naturelle*, ed. Schmitt and Crémière, 'Introduction', vol.1, p.33-34.
22. Buffon, 'Essais d'arithmétique morale', p.56-57.

causes of nature.[23] Within this process, space was left for the possible and probable – degrees of imagination which had the potential to become believable and hence true, 'de simples faits qu'il faut mettre en réserve, en attendant qu'un plus grand nombre d'observations & une plus longue expérience nous apprennent d'autres faits & nous découvrent la cause physique, c'est-à-dire, l'effet général dont ces effets particuliers dérivent' ('simple facts that we must store up, waiting until a greater number of observations & more experience teach us other facts & reveal to us the physical cause, that is, the general effect from which these particular effects derive').[24] This process could be as indefinite as the movement and progress of time. Yet, as they acquired more general knowledge of nature, people could learn to play an increasingly active role in the compilation and narration of nature's story. A hypothesis was a tool that pushed the intellect forward onto seemingly shaky epistemological ground, asking readers to use their current knowledge in order to judge a range of probable causes, and to decide – based on the inner light of reason and certainty – which was most likely to happen next. In this decision, the rational order of the mind aspired to the physical order of nature, hoping to act in its place.

Discovering the true laws of nature and what really happened at the beginning of the universe: these goals were less important to Buffon than presenting a system whose details were almost all readily accepted by the intellectual community, with just enough probabilistic space to make intellectuals comfortable in the exploration and the invention of the unknown through reason and language. The hypothesis of the comet in the *Histoire naturelle* was indicative not of a permanent explanation, but of the natural activities yet to be discovered by the historian, as of the relationships yet to be forged between the mind and the world. The text of natural history was a monument attesting to the current extent of human knowledge. Its style showed the degree to which Buffon thought people understood nature in the 1750s. The probable, a transitional and discursive space between reason and reality, was a domain for conjecture, asking readers whether the logical order of their thoughts was proportional *enough* to the physical order of the world so as to be capable of inventing a convincing discourse of what could not yet be perceived. Using hypothesis, Buffon provided his readers with the rational and rhetorical tools to continue investigating and learning about the world, modifying the story of

23. Buffon, 'Essais d'arithmétique morale', p.57.
24. Buffon, 'Essais d'arithmétique morale', p.57-58.

nature as their perception of it changed. As Benoît de Baere argues, the purpose of invention in the *Histoire naturelle* was to allow readers to gain confidence in Buffon's voice as it guided the observation and ordering of the world and showed how to build analogies and *rapports*. This first, methodical stage of invention was important because, when Buffon's readers reached the limits of the visible world and had to delve into the no-longer visible realm of the deep past, represented by Buffon's poetic language, he wanted them to continue to reason in the same manner. At the moment of the hypothesis, methods of scientific induction and methods of discursive invention merged.[25] Indeed, they were difficult to distinguish in Buffon's text because he intended to make them appear discoverable and legible in the same manner.

Hypothesis and heat: inventing the hidden mechanism of nature

Alongside the foundational plan of style in the *Histoire naturelle*, hypothesis was the starting point from which to build and order a body of knowledge, and to comprehend and experience the underlying mechanism of nature as it created over time. As de Baere explains, one of the purposes of hypothetical invention in the *Histoire naturelle* was 'l'élaboration de connaissances dont l'utilité est moins immédiate – celles qui nous font entrevoir la "marche" de la nature' ('the elaboration of knowledge of which the utility is less evident – that knowledge which allows us to see the "march" of nature').[26] Ultimately, Buffon's hypothesis was a way to see differently in order to attain knowledge of nature's hidden workings. If presented through a language that captured nature's vitality and movement, it could not only be justified through reason and probability, but it could also convince through feeling. It was therefore another form of writing with 'heat'.

To recall Buffon's hypothesis from the 'Preuves de la théorie de la terre' mentioned at the beginning of this chapter: far back in the prehistory of the solar system, a comet had smashed into the Sun, causing small pieces of this star to break away from it. These pieces were to become the planets of the modern solar system.[27] Buffon began this essay with a presentation of contemporary knowledge about

25. De Baere, *La Pensée cosmogonique de Buffon*, p.153-55.
26. De Baere, *La Pensée cosmogonique de Buffon*, p.127.
27. Buffon, *Histoire naturelle* (1749-1767), vol.1, p.133.

the Earth and the other planets, taking care to remind his readers that this understanding was the result of a consensus on the significance of previous natural historical discoveries, as well as part of the continuous process of knowledge production.[28] After all, the currently accepted facts were once themselves a collection of contentious, possible or probable claims, but they had acquired the more certain status of knowledge as each had contributed to the comprehension of the others. Buffon explained how, in the progress and history of astronomy from Galileo to Newton, one idea was built upon another, forming a contingent chain that documented the history of the development of present knowledge of the laws of planetary motion.[29]

The choice of a comet as the celestial phenomenon which precipitated the beginning of the solar system probably emerged from a general usage of comets at the time as a trope for impending change or disaster. It perhaps transpired from Buffon's reading of Newton's discussion in the *Principia mathematica* of the comet of 1680, which came close to the Sun and which Newton believed had the potential to perhaps hit the Sun in its next revolution.[30] Buffon cited Newton on this point, but used provocative language to push the limits of the Newtonian epistemology of hypothesis:

> Pour peu qu'on examine le cours des comètes, on se persuadera aisément qu'il est presque nécessaire qu'il en tombe quelquefois dans le soleil. Celle de 1680 en approcha de si près, qu'à son périhélie elle n'en étoit pas éloignée de la sixième partie du diamètre solaire; & si elle revient, comme il y a apparence, en l'année 2255, elle pourroit bien tomber cette fois dans le soleil; cela dépend des rencontres qu'elle aura faites sur sa route, & du retardement qu'elle a souffert en passant dans l'atmosphère du soleil. *Voyez Newton, 3e édit. pag. 525.* Nous pouvons donc présumer avec le Philosophe que nous venons de citer, qu'il tombe quelquefois des comètes sur le soleil.

28. Buffon, *Histoire naturelle* (1749-1767), vol.1, p.127-28.
29. Buffon, *Histoire naturelle* (1749-1767), vol.1, p.129-30.
30. Buffon was not the only natural philosopher to attribute the origin of Earth and the planets to a comet: William Whiston, a disciple of Newton in the tradition of sacred physics (correlating natural history and biblical events), began his 1696 *New theory of the Earth* not with divine intervention but with a comet. See Paolo Casini, 'Buffon et Newton', in *Buffon 88*, ed. J.-C. Beaune *et al.*, p.299-308 (305). On Whiston's work, see Norman Cohn, *Noah's flood: the Genesis story in western thought* (New Haven, CT, 1996), p.62-71; and Jean Seidengart, 'Le traitement du problème cosmologique dans l'œuvre de Buffon', in *Buffon 88*, ed. J.-C. Beaune *et al.*, p.309-26 (311-13).

> If we study the orbit of comets, we will easily persuade ourselves that it is almost necessary that some [comets] will occasionally fall into the Sun. The comet of 1680 came so close [to the Sun], that at its perihelion it was no farther away than one sixth of the Sun's diameter; and if it returns, as it seems it will do in the year 2255, it could this time really fall into the Sun. That depends upon what it would have encountered in its path, and upon the delay it would have suffered passing into the atmosphere of the Sun. *See Newton, third edition, page 525.* We can therefore presume with the philosopher we just cited, that sometimes comets fall into the Sun.[31]

The problem of explaining the origins of the universe was troubling to Newton, particularly so because of the existence of comets. It was difficult to accept them as purely natural objects, even though they had been shown by Edmund Halley to follow a regular, though highly elliptical, orbit. Alongside their older, supernatural meaning as signs of divine intervention and harbingers of natural disasters, plagues or regime changes, comets had been given a mathematical explanation and their movements could be reliably predicted. Halley's work allowed comets to be understood as part of the natural system of the universe, governed by its laws. Nevertheless, their presence continued to call the harmony, beauty and unity of God's creation into question. For instance, they did not move in the same direction and plane around the Sun as the planets. Newton believed that, as he wrote in his *General scholium*, although people had the ability to figure out God's natural laws through the mathematical calculation and abstraction of the regular behavior of natural phenomena, they could never grasp the totality of the divine plan. They could only understand a sliver of the universe through reason; the rest was to be accepted through the faith required of the subjects of God's 'dominion'. Comets fell somewhere in between reasonable research and comprehension and faithful compliance. De Baere notes that, according to Newton, after a certain point the entirety of the natural world was not meant to be comprehended by humans: 'l'ordre et la régularité sont des attributs de la divinité' ('order and regularity are attributes of the divine') and 'l'ordre planétaire a une origine surnaturelle et, par conséquent, inaccessible au savoir humain' ('the planetary order has a supernatural origin, consequently inaccessible to human knowledge').[32]

31. Buffon, *Histoire naturelle* (1749-1767), vol.1, p.135.
32. De Baere, *La Pensée cosmogonique de Buffon*, p.49.

As Jean Seidengart explains, Buffon, like other French philosophers of the time, was dissatisfied with what he saw as Newton's recourse '[à] Dieu ou au hasard' ('[to] God or to chance') as ultimate explanations for the universe.[33] Instead of rejecting the Newtonian order of nature, however, Buffon built his hypothesis on what he considered to be the unused space that Newton claimed was inaccessible to human knowledge: between what was known from observation of celestial phenomena and what it seemed that only God could know. In this space, his work implied, lay the hidden plan of all natural activity, the manner or hidden mechanism through which nature created and articulated its phenomena. This mechanism was discoverable through conjectural questions Buffon had not found appropriate to ask earlier in the *Histoire naturelle*, because they were not about the direct observation of nature but rather about the phenomena of the past that fell outside of the scope of what people could perceive visually – phenomena that, for Buffon, had to be presented through the language of heat.

Recalling the discussion in Chapter 2, heat could be understood as a new way of seeing, provoked by a change or event. Buffon's language narrated two types of event: one which initially happened in the world, and then was subsumed and harmonized into the continuous, cyclical flow of the story of natural history, and the other which initially occurred in the imagination. The event produced by imagination created shock or astonishment because it had not yet been assimilated into accepted knowledge. This second sort of event was a form of rupture, and the sudden incident of a comet colliding with the Sun may be read in this manner. It suggested change, a spark or break in the steady spiral-like advancement of historical time. This break was necessary to the function and advancement of both nature and science. To style's continuous regularity, hypothesis was the force of change.

Buffon understood the history of the world to advance through a play between continuity and rupture. Drawing on Newton's *Principia mathematica* and 'Queries' from the *Opticks*, Buffon explained that all the movement in the solar system was the result of two forces. The first is 'une attraction du haut en bas, ou de bas en haut vers un centre' ('an attraction from the top to the bottom, or from the bottom to the top, toward a center'): the motion of gravity, which pulled particles

33. Seidengart, 'Le traitement du problème cosmologique dans l'œuvre de Buffon', p.314.

in toward one another. The second type of movement was 'une impulsion de droite à gauche, ou de gauche à droite' ('an impulsion from right to left, or from left to right'), that is, a linear displacement resulting from an initial impulse produced by contact with another moving body. The comet crashed into the Sun because of attraction, but, as soon as these two bodies collided, parts of the Sun were driven away from it by the countering force of impulsion. Attraction then came into play again, acting on the lateral displacement of the initial burst of impulsion, consolidating and shaping the fragments of the Sun into spherical planets, and holding them in regular, elliptical orbits around their star. Since attraction gave the planets their form, it could also be used to account for the shape of the Earth and for many of the regular physical phenomena and events that occurred on it over the course of time.[34] Impulsion, by contrast, could help to explain the plane and direction of movement of the planets. Moreover, since it was known that the planets all moved in the same direction around the Sun, Buffon inferred that they were all created by the same change or shock: 'cette conformité de position & de direction dans le mouvement des planètes, suppose nécessairement quelque chose de commun dans leur mouvement d'impulsion, & doit faire soupçonner qu'il leur a été communiqué par une seule & même cause' ('the conformity of position & direction in the movement of the planets necessarily implies something shared in their impulsive movement, and makes us wonder whether it [this movement] was not communicated to them by one and the same cause').[35] Thus, rather than its origins being explained as part of the mystery of God's creation, the current workings of the solar system attested to the interaction in the past of a singular event, the impulsive shock of the comet, with a continual force, the regular state of gravity.

In accordance with his definition of style as the harmonization of the natural historian's story to the order and activity of nature through the heat of his prose, in Buffon's hypothesis the structure of his argument should have been in line with how he had already explained history in the 'Second discours'. Yet the singular, isolated event of the comet appeared to contradict this harmony as well as the conception of natural history as the narrative of the cyclical, repetitive processes of the *longue durée*. The sudden, impulsive movement of the comet, in contrast with gravity's perpetual, steady work of pulling

34. Buffon, *Histoire naturelle* (1749-1767), vol.1, p.128.
35. Buffon, *Histoire naturelle* (1749-1767), vol.1, p.133.

bodies together and contradicting movement, was perhaps designed by Buffon to be felt by readers before it could be rationally comprehended. He worked to ensure that the initial shock of his argument did not appear to be either a random, meaningless event, or an attempt to dazzle readers' imaginations. It was as if he strove to capture the energy created by the comet and make it endure in readers' minds, encouraging them to remain engaged and continue to read about the ensuing history of the development of the solar system. Having struck with a linguistic force equivalent to that of the comet, he then began to insert the event into a larger chain of reasoning and history. Taking into account contemporary knowledge about the laws of motion, Buffon asserted:

> Ne peut-on pas imaginer avec quelque sorte de vrai-semblance, qu'une comète tombant sur la surface du soleil, aura déplacé cet astre, & qu'elle en aura séparé quelques petites parties auxquelles elle aura communiqué un mouvement d'impulsion dans le même sens & par un même choc, en sorte que les planètes auroient autrefois appartenu au corps du soleil, & qu'elles en auroient été détachées par une force impulsive commune à toutes, qu'elles conservent encore aujourd'hui?
>
> Could one not imagine, with some degree of verisimilitude, that a comet, falling upon the surface of the sun, would have moved this star, and that it would have broken off a few little pieces [from the sun] to which it would have communicated an impulsive movement, in the same direction and from the same force of impact, such that the planets would have formerly belonged to the body of the sun, and would have been detached from it by a shared impulsive force that they still retain today?[36]

By asking his readers to try to 'imagine, with some degree of verisimilitude', Buffon placed them in the mindset of Cartesian thought. He pushed them to wonder, 'what would happen if…', and to follow the ensuing chain of cause and effect to rational, although not observable, conclusions. Unlike the vital energy or heat developed by Buffon's description of the horse, here heat lacks a current, analogous model in nature, and is derived almost entirely from the power of his language. In a certain sense it did not matter whether a real collision between a comet and the Sun created the solar system. Rather, the story formed by working from the comet outwards to its effects had 'verisimilitude'.

36. Buffon, *Histoire naturelle* (1749-1767), vol.1, p.133.

It conformed to and justified the perceptions of the universe held by contemporary astronomers, who could observe and verify with their own eyes (albeit through a telescope) that the planets moved similarly, and that this might be a sign of a common origin aptly explained by the collision event of a comet.

In Buffon's hypothesis, the fact that harmony was interrupted and then reestablished was important. Just as he thought that the force of attraction reimposed regular motion in the universe, smoothing out the bumps of impulsion, style reinforced proportion in Buffon's argument. His reasoning in the 'Preuves' slowly reinserted the comet into his system of the world, both as a historical event and as a stylistic gesture. This event, which happened with explosive power (and, most likely, great amounts of actual heat) at the beginning of the universe, was gradually tempered to the present state of the world, as to the present state of human knowledge of the world. In the 'Preuves', Buffon's readers witnessed the work of style in creating harmony between physical and mental worlds. They experienced how the hypothesis unfolded in a narrative, advancing in time from the initial crash of the comet into the sun to the present state of the universe. Buffon even entered into an imaginary dialogue with his ideal readers, a training process that he thought would gradually make them more capable of reading and interpreting the story. For instance, Buffon frequently paused to relate the elements of the story to current knowledge, rendering it more believable to his interlocutors. He adopted two voices: his own, 'je', the specialist who has conjectured back into time, and the staged voice of the reader, 'on', currently on the outside, shocked and confused by the hypothesis. 'On' frequently interrupted, asking questions that allowed Buffon to both clarify the hypothesis and move the story forward. Through this gesture, he asked his reader to watch how, from the initial shock received when encountering the hypothesis for the first time, he or she might become reintegrated into the belief system of the *Histoire naturelle* through the metaphors of rupture and regularity embodied in the relationship between impulsion and attraction. As regularity had been restored to the physical world through gravitational force, so harmony, between Buffon's argument and the actual regularity of the world, was consequently restored to the *Histoire naturelle*. Each time Buffon introduced his readers to an aspect of the hypothesis, the two voices, 'je' and 'on', converged into one pronoun, the first-person plural 'nous'. The imagined reader asked – 'But, one would say' ('Mais, dira-t-on') – why, since the matter that formed the planets emerged from the Sun, did the orbits of the planets not graze the surface of this star each time they passed by it? Buffon then explained – 'To this, I reply' ('A cela je réponds') – how the planets were pushed far from the Sun, through an acceleration caused by the force of the comet's shock. He interwove this elucidation with practical examples and current observations of similar types of movement, such as the ejection of material from the eruption of Mount Vesuvius earlier in the century, or the trajectory of an imagined bullet or rocket ('une fusée volante').[37] After presenting these explanations, Buffon envisioned an agreement between himself and his readers, and the third, harmonious voice emerged, placing the two parties on the same level of belief and desire to learn: '*suivons*-en l'objet [de l'hypothèse] & *tirons* des inductions; *voyons* donc ce qui a pû arriver lorsque les planètes [...] ont reçu ce mouvement d'impulsion' ('*let us follow* the object [of the hypothesis] & induce; *let us see* what could have happened when the planets [...] received this movement of impulsion').[38]

The hypothesis of the formation of the planets did not persuade readers by providing them with a definitive answer of how and why the Earth came to have its present shape and characteristics. Rather, it generated a process of thinking about the history of the world through the energy of style. The hypothesis of the *Histoire naturelle* and the strategies it imparted did not so much offer a way of describing nature – for this would have been too static and too opposed to the underlying movement and change of nature through time, as to the sudden shocks it supposedly endured from time to time. Instead, they showed how to make nature understandable and legible, and how to communicate the dynamic 'heat' which brought the world into existence and motion. The discovery of nature's past was thus rendered possible and comprehensible through the discursive activity of inventing this past in the form of persuasive argument.

37. Buffon, *Histoire naturelle* (1749-1767), vol.1, p.139-43. Seidengart notes that Buffon wrote an account in 1740 of rockets that he had built, entitled *Mémoire sur les fusées volantes*. He explains, however, that the calculations about trajectory movement that Buffon attached to the proof of his hypothesis were considered insufficient and not convincing enough by contemporaries such as D'Alembert. See Seidengart, 'Le traitement du problème cosmologique dans l'œuvre de Buffon', p.318-19.

38. Buffon, *Histoire naturelle* (1749-1767), vol.1, p.143 (my emphasis).

Making heat real: the hypothesis in the 'Epoques de la nature'

Buffon returned to the same hypothesis of the formation of the planets about thirty years later in the 'Epoques de la nature'.[39] In this work, however, heat was no longer simply a stylistic tool – in other words, by the time of the 'Epoques', Buffon seemed to think that his language had become synonymous with the natural world. The hypothesis was the basis for understanding the reality of the present-day globe. Organized in seven chapters, or 'epochs,' the 'Epoques' told the chronological story of the Earth from its beginning, when the planetary system came into existence by collision with a comet after this comet had collided with the fire of the Sun.[40] Fragments of the Sun became the planets; thus, like the other planets, the Earth began as a molten globe. Its story could be written with literary heat *because* it was about how heat had formed the planet. Furthermore, the next chapter will show that, as the world slowly lost its heat over time, Buffon's narrative gained energy – and it was this heat that he sought to transmit to readers.

Over the course of the 'Epoques', Buffon traced the history of the planet Earth through the gradual cooling of the temperature of the globe, from its fiery origins to its current state. Clues in current astronomical and geological observations – the movements of the planets and comets, for instance, or the formation of vitric matter into sedimentary layers of rock – became proofs that led back, chronologically and rhetorically, to the original impulsion that caused the comet to break off a piece of the Sun. All events in his story hinged on the plausibility of the hypothesis of the comet, and the hypothesis in turn justified currently accepted facts such as the shape and rotation of the Earth (which, as I will explain, could not have attained its current state without the action of fire), becoming the foundational plan that

39. The 'Epoques' were published in the fifth volume of the *Supplément à l'Histoire naturelle*, in 1778. The work did not reach the general public for purchase until April 10, 1779. See Roger, *Buffon*, p.554. An English translation of the 'Epoques' is forthcoming: Buffon, *The Epochs of nature*, ed. and translated by Anne-Sophie Milon, Jan Zalasiewicz and Mateusz Zalasiewicz (Chicago, IL, 2018).

40. According to Roger, concluding the 'Epoques' in six, rather than seven, chapters was Buffon's initial intention. See Roger, 'Introduction', in *Les Epoques de la nature*, ed. Jacques Roger, Mémoires du Muséum national d'histoire naturelle, series C, vol.10 (Paris, 1962), p.ix-clii (xxxv).

held both the seeds of historical development and the development of Buffon's argument.[41]

Buffon began to give probable weight to his hypothesis through the analysis of accepted facts about the Earth. For him, these facts generated a particular form of argumentative proof concerning the events of the Earth's past, and especially those related to heat. For example, Buffon wrote about the shape of the terrestrial globe, as determined by Clairaut's and Maupertuis's voyage to Lapland earlier in the century to measure an arc of meridian. The physical observations and mathematical calculations of these philosophers had shown that the Earth was not a perfect sphere but rather slightly flattened at the poles and slightly rounded at the equator. This fact became a step in Buffon's proof that the planet Earth began in fire:

> Ainsi la première conséquence qui sort de ce fait incontestable, c'est que la matière dont notre terre est composée étoit dans un état de fluidité au moment qu'elle a pris sa forme, et ce moment est celui où elle a commencé à tourner sur elle-même. Car si la terre n'eût pas été fluide, et qu'elle eût la même consistence que nous lui voyons aujourd'hui, il est évident que cette matière consistante et solide n'auroit pas obéi à la loi de la force centrifuge, et que [...] la terre, au lieu d'être un sphéroïde renflé sur l'équateur et aplati sous les pôles, seroit au contraire une sphere exacte, et qu'elle n'auroit jamais pu prendre d'autre figure que celle d'un globe parfait.

> Thus the first consequence that can be derived from this indisputable fact is that the matter of which our Earth is composed was in a state of fluidity when it took shape, and this is the moment where it began to rotate upon itself. For if the Earth had not been fluid, and had the same consistency as what we observe today, it is evident that this solid and heavy matter could not have obeyed the law of centrifugal force, and that [...] the Earth, instead of being a sphere-like body swollen at the equator and flattened at the poles, would be on the contrary a true sphere, and could have never taken any other shape than that of a perfect globe.[42]

Buffon used contemporary knowledge of the Earth's shape to create

41. On the imagination, construction and ordering of the 'epoch' chapters of the 'Epoques', see de Baere, *La Pensée cosmogonique de Buffon*, p.241-42, and Kenneth Taylor, 'Buffon, Desmarest, and the ordering of geological events in *époques*', in *The Age of the Earth: from 4004* BC *to* AD *2002*, ed. C. L. Lewis and S. J. Knell (London, 2001), p.39-49 (42).
42. Buffon, 'Des époques de la nature', p.7.

a comparison that allowed readers to reason about a moment and a state that they never could have physically observed. His conclusion was based upon the degree to which he thought his public was already persuaded by the original fact. If the readers were quite certain about the shape of the Earth, then they had to at least accept as probable an explanation of how it came to be this way. Not only did the fact of the shape of the Earth render Buffon's conclusion more plausible, but his argument in this passage implied that if his inference were different or to the contrary – 'if the Earth had not been fluid' – then the present shape of the Earth would also be different. The natural history of the planet became contingent upon the rhetorical and discursive power of an invented story. Facts were based upon the discovery of an underlying structure, a hypothesis which stated that the Earth had begun as fire and that heat had written its story.

Buffon continued to draw conclusions from the assertion that the Earth, to attain its present shape, had to have first been composed of a type of liquid. A liquid state of matter implied some kind of heat, or change of physical state, at work, from whence Buffon gathered that the Earth could have been a sphere of either water or fire at its beginning. Considering the current state of the globe, he added another fact: that more of the Earth was composed of 'arid matter' ('matière aride') than of water. There was not enough water to dissolve all this dry matter. He thus concluded that 'cet état primitif de fluidité dans le globe terrestre' ('this primitive state of fluidity of the terrestrial globe') was one of fire.[43] The explanation increased in believability as it was compared to two other current facts. First, Buffon held that the Earth had 'une chaleur intérieure qui lui est propre' ('its own inner heat'), which did not come from the Sun. Second, the heat the Earth received from the Sun 'est assez petite, en comparaison de la chaleur propre du globe terrestre' ('is fairly small, in comparison with the terrestrial globe's own heat').[44] The first fact, '[c]ette juste conséquence déjà très-vraisemblable par elle-même' ('this just consequence, already quite believable on its own'), explained that the original liquid state of the Earth was igneous, and

43. Buffon, 'Des époques de la nature', p.7-8.

44. Buffon, 'Des époques de la nature', p.5-6. Buffon took these facts from the work of a colleague at the Académie des sciences, Dortous de Mairan: 'Nouvelles recherches sur la cause générale du chaud en été et du froid en hiver' (1765), which proposed that the Earth was mostly heated by an internal source, not the Sun. The significance of this work will be discussed in greater detail in the next chapter.

the second fact rendered this statement 'a certainty'.[45] Buffon then added a third fact, that the rocky substances that comprised the planet Earth 'sont en général de la nature du verre, et peuvent être toutes réduites en verre' ('are generally of the nature of glass, and all can be reduced to glass').[46] By this he meant that the minerals found in the Earth's strata had vitrifiable tendencies and could ultimately be melted back into the liquid and fiery initial state of the planet.

To further reconstruct the history of nature's dispensation of heat, Buffon explained in the introduction to the 'Epoques' that 'il faut fouiller les archives du monde, tirer des entrailles de la terre les vieux monumens, recueillir leurs débris' ('it is necessary to search the archives of the world, extract ancient monuments from the entrails of the earth, collect their debris').[47] The monuments Buffon chose were examples of what he called 'debris' surviving from the past, namely fossils: pieces of shells and residue of other marine life, or the 'dépouilles' ('remains') of bones and tusks from large mammals such as elephants, rhinoceroses and hippopotamuses. The remnants of these organisms were found in layers of limestone and other 'matières calcaires' ('chalky matter'), strata that Buffon believed formed later in the Earth's history because they did not contain as much vitrifiable material as those further down and closer to the Earth's center. Besides their fragmentary and ruined nature (Buffon described shards of bone and shell, often so completely disintegrated that they melded and blended with the surrounding stone), these monuments were also intriguing precisely because they were found at different depths of the Earth's crust – the shells deeper in the ground than the elephant bones, for example – and in unexpected places – the elephant bones in northern countries, the shells in hills and mountains far inland.[48] The type of fossilized monument and its location, both in geographical space and within the sedimentary rock layers, allowed Buffon to divide the historical space of his narrative into more defined periods, piecing together what happened in each period, and showing how each epoch was distinct from the others. In this way, what was initially 'debris' or ruins became a 'monument' attesting to prior events.

Recalling the role of monuments in style, discussed in Chapter 3, they were the coproduction of nature and the natural historian:

45. Buffon, 'Des époques de la nature', p.8.
46. Buffon, 'Des époques de la nature', p.6.
47. Buffon, 'Des époques de la nature', p.1.
48. Buffon, 'Des époques de la nature', p.15-16.

Buffon noticed the potential energy within seemingly random debris and used language to reconstruct its story and significance. For example, an elephant tusk found in the superficial sedimentary layers of a northern country would continue to belong to the vast, ruined archives of the world, until it was discovered and then interpreted with respect to the change of temperature over time. Applying this rule to the remnants of the past allowed the natural historian to read each rock layer in which the monuments were contained as indicative of a cooler temperature than the layer below it, and then to assemble these remnants in a chronological and narratological order. Each monument was transformed into a chapter heading or an epoch in Buffon's history of nature. Monuments were the product of the vivid language of the writer of natural history, who identified, decoded and reconstructed seeming ruins, returning to them the natural energy they once had. This discursive heat allowed the reader to feel as if she or he were directly observing the formation of mountains, the explosion of volcanoes, the advance and retreat of the oceans.

Starting with the initial period of molten rock, the natural history of the Earth in the 'Epoques' was one of gradual cooling. Through monuments, Buffon imagined and reanimated the activities that would have been feasible at a specific temperature and time. He metaphorically dug into layers of rock to revive in words the events of the distant past. In the 'Second epoch', for example, subtitled 'Lorsque la matière s'étant consolidée a formé la roche intérieure du globe, ainsi que les grandes masses vitriscibles qui sont à sa surface', he wrote:

> Représentons-nous, s'il est possible, l'aspect qu'offroit la Terre à cette seconde époque, c'est-à-dire, immédiatement après que sa surface eut pris de la consistence, et avant que la grande chaleur permît à l'eau d'y séjourner ni même de tomber de l'atmosphère [...]. Qu'on se figure pour un instant la surface actuelle du globe dépouillée de toutes ses mers, de toutes ses collines calcaires, ainsi que de toutes ses couches horizontales de pierre, de craie, de tuf, de terre végétale, d'argile, en un mot de toutes les matières liquides ou solides qui ont été formées ou déposées par les eaux; quelle seroit cette surface après l'enlèvement de ces immenses déblais? Il ne resteroit que le squelette de la Terre, c'est-à-dire, la roche vitrescible qui en constitue la masse intérieure.
>
> Let us imagine, if it is possible, the appearance offered by the Earth in this second epoch, that is, immediately after its surface had solidified, and before the great heat had allowed water to remain on the surface or fall from the atmosphere [...]. Imagine for a moment the actual surface of the globe stripped of all its seas, of all its limestone hills,

> as well as of all its horizontal strata of rock, chalk, tuff, humus, clay – in sum, of all liquid and solid matter that was formed or deposited by water. What would this surface be after removing this immense rubble? There would only remain the skeleton of the Earth, that is, the vitrifiable rock that constitutes its interior mass.[49]

By way of imagination, the readers of the 'Epoques' became witnesses in this thought experiment to phenomena they neither could have observed with the naked eye, nor could have conceived without the construal of temperature and time that structured Buffon's history. Using the monument of the second epoch, 'les plus hautes montagnes composées de matières vitrescibles, [...] les seuls témoins de cet ancien état' ('the highest mountains made up of vitrifiable matter, [...] the only witnesses of that ancient period'), Buffon peeled away the layers of rock, time and temperature, revealing 'the skeleton of the Earth'.[50] Upon this skeleton, as upon the original hidden plan of style, he rebuilt the world in language.

Each subsequent chapter of the 'Epoques' moved farther away from the original source of heat, but continued to be legible through this starting point. In the 'Troisième époque: lorsque les eaux ont couvert nos continens', the remains of shells found in limestone several layers higher than the volcanic rock were described.[51] Buffon assumed that the shells were of a later period than that of the original fire. Because they resembled modern marine species, they were most likely part of an early ocean. The fossils were found inland, far from the present-day ocean, so the historical ocean must have been vast, covering the modern continents. If an ocean had existed, then the Earth must have been cool enough to support liquid water without boiling and evaporating it away. This permitted the existence of life forms nourished by water, yet larger and even more prolific than present marine life, because formed at a time when it was much warmer and when nature did not yet have to economize heat.[52]

Even as the background energy from the beginning of the Earth's creation slowly diminished, nature sometimes had local bursts of

49. Buffon, 'Des époques de la nature', p.76-77. The translation of the title of the 'Seconde époque' is 'When matter, having consolidated, formed the interior rocks of the globe, as well as the great vitrifiable masses on its surface'.
50. Buffon, 'Des époques de la nature', p.82.
51. Buffon, 'Third epoch: when the waters covered our continents', 'Des époques de la nature', p.98-100.
52. Buffon, 'Des époques de la nature', p.17, 24-25.

creative power. In the 'Quatrième époque', for instance, the waters of the ocean retreated and volcanoes spewed a fresh layer of lava on top of the limestone. This added heat paved the way for a fifth epoch, during which large mammals thrived. The monuments of their bones were found in strata closer to the surface of the Earth and thus closer in time to the present day; these fossils were also located in northern regions of the globe. Elephants, which currently lived in warm climates close to the equator, had been able to live well above the equator (or at least this is what the location of the fossilized remains of mammoths evidenced to Buffon), indicating that the world must have been warmer in the past.

As Buffon ordered and resuscitated the activities of the past, the epistemological value and creative potential of his original hypothesis grew stronger. In his presentation of the hypothesis of the comet in the first chapter of the 'Epoques', Buffon described it in the conditional, probable, tense. He implied that what was now viewed as a hypothesis might someday be read as a very reasonable explanation for the current state, order and movement of the solar system. By placing the hypothesis at the historical beginning of the world, and at the rhetorical beginning of his story of nature, Buffon set up two threads whose ontological and narrative distinctions had already begun to converge:

> Ce n'est donc pas que j'aie affirmé ni même positivement prétendu que notre Terre et les Planètes aient été formées nécessairement et réellement par le choc d'une comète, qui a projeté hors du Soleil la six cent cinquantième partie de sa masse: Mais ce que j'ai voulu faire entendre, et ce que je maintiens encore comme hypothèse très-probable, c'est qu'une comète qui, dans son périhélie, *approcheroit* assez près du Soleil pour en effleurer et sillonner la surface, *pourroit* produire de pareils effets, et qu'il n'est pas impossible qu'il se forme quelque jour de cette même manière des planètes nouvelles, qui toutes circuleroient ensemble comme les planètes actuelles, dans le même sens, et presque dans un même plan autour du Soleil [...]: en sorte que tous les phénomènes de ces planètes possibles et idéales, *seroient* (je ne dis pas les mêmes), mais dans le même ordre, et dans des rapports semblables à ceux des phénomènes des planètes réelles.
>
> It is not, therefore, that I affirmed or even definitively claimed that our Earth and the planets were necessarily and really formed by the impact of a comet which ejected from the Sun 1/650th of its mass. But what I wanted to make understood, and what I still maintain as a very probable hypothesis, is that a comet which, in its perihelion,

> *would come* close enough to the Sun to graze and cross its surface, *could* produce the same effects, and that it is not impossible that someday new planets form in the same manner, all revolving together in the same direction and on the same plane around the Sun as the actual planets, such that all the characteristics of these possible and ideal planets *could be* (I do not say the same [as the current planets]) but in the same order and with similar relationships as those of the characteristics of the real planets.[53]

In the 'Epoques', as I have shown, Buffon created a world in thought, structured by the hypothesis of the comet, which established a framework of legibility for current facts and monuments. In the above passage, he described his story as 'very probable', using the conditional tense to distinguish it from what happened 'necessarily' and 'really' in nature. He did not 'affirm' or 'claim' his account to be true, but instead appealed to the intellect of the reader, to make him or her understand by entering into conditional mode and considering the probability of the argument. The comet of Buffon's thought experiment could come close enough to the Sun, and thus 'could produce the same effects' as those observed in the real solar system today. Like the invention of a new world in Decartes's *Le Monde*, the story of the 'characteristics of these possible and ideal planets' followed the same principles as the 'characteristics of the real planets'.

As the story of the 'Epoques' moved forward in time, the hypothesis became no longer simply a possible account of the Earth's origin, but rather the crucial point from which the plot of the story of nature emerged. It ordered and made nature comprehensible through heat. Without it, there would be neither order nor temporal legibility; readers would be able neither to 'remonter aux différens âges de la Nature' ('go back to the different ages of Nature') nor to understand the present state of the world.[54] From the beginning of the 'Epoques', there was already a hint that the reading experience would eventually render the probability and believability of the hypothesis inseparable from the reality of the world. Indeed Buffon proposed to reenact the epochs, the great movements in the story of natural history, 'telles qu'elles sont ou telles qu'elles pourroient être: car ces deux points de vue sont à-peu-près les mêmes' ('as they are or as they could be: for these two points of view are practically the same').[55] The events

53. Buffon, 'Des époques de la nature', p.53-54 (my emphasis).
54. Buffon, 'Des époques de la nature', p.1.
55. Buffon, 'Des époques de la nature', p.53.

of the world and the story of these events would become ever more legible through the same principles and practices: as nature created with heat, so Buffon composed with heat. It was crucial that readers understand the past and present as points in a continuous narrative structured through the dispensing of heat for, as I will argue in the following chapter, it was through this perspective that they might gain the ability to hypothesize upon the future of the world. From the thought experiment of nature's past, they could learn to invent the heat nature could no longer produce on its own, and to use the world as a laboratory for the creation of new experiments whose energy had the potential to change reality.

5. Reinventing nature's heat

La seule expérience est un guide pour moi;
Instruire est son devoir, et peindre est mon emploi;
Mes pinceaux sont trempés, et la vive lumière
Dans mes riches tableaux brillera la première,
La lumière, inconnue en ses secrets ressorts,
Qui frappe, échauffe, éclaire et pénètre les corps,
Donne à l'air respiré sa pureté vitale,
Aux plantes organisées leur âme végétale,
Epanche ses torrents de la hauteur des airs,
Au centre de la terre, aux profondeurs des mers,
Inonde incessemment des régions sans nombre,
Et, traversant d'un trait les royaumes de l'ombre,
Du trône ardent du jour prend un essor pareil,
Au coup-d'œil de ce Dieu qui créa le soleil;
De bienfaits, de beautés source immense et féconde;
Enfin, l'âme, la vie et le peintre du monde.

Experience alone is a guide for me;
To instruct is its duty, to paint is my task;
My brushes are dipped, and the lively light
In my rich paintings will be the first to shine,
Light, unknown in its secretive motives,
That strikes, heats, illuminates and penetrates bodies,
Gives to the breathed air its vital purity,
To organized plants their vegetal soul,
Pours forth its torrents from the height of the heavens,
From the center of the earth, from the depths of the seas,
Incessantly floods innumerable regions,
And, crossing in a flash the kingdoms of shadow,
From the ardent throne of the day takes a similar flight,
To the glance of that God who made the sun;

> Of beneficence, of beauties [light is] the immense and fertile source;
> And finally the soul, the life, and the painter of the world.[1]

The manner in which Buffon construed the march of natural history in 1749, at the outset of the *Histoire naturelle* project, was very different from the manner in which he did so in the 1778 text, the 'Epoques de la nature'. In the first volume of Buffon's work, as discussed in Chapter 3, the Earth was construed as having been shaped over time by water; in the 'Epoques', he instead ascribed the Earth's formation to fire. In both versions, he used the same hypothesis of the comet to explain the planet's origin.[2] The building up and eroding away of continents by water described in 1749 was a largely repetitive, cyclical process that had gone on for many years prior and that would continue to happen into the future. In 1778, however, the history of the Earth had a definite beginning and end. Moreover, the hypothesis of the comet and the resulting story of its effects were proposed much less tentatively.[3] According to Buffon, it was now very probable that the Earth originated as a fiery rock torn from the Sun by a comet. The residual heat from this event gave nature much of its productive power, yet this power had consistently been waning as the Earth aged and cooled. Over the eras of geological time, Buffon believed, the internal heat of the planet had dissipated and had been slowly replaced by inactivity and coldness, in such a way that it could be reasonably foreseen that the world would someday end in ice. At some point in the future, nature would no longer be able to continue to write its history on its own. As Jacques Roger describes the 'Epoques':

> [L]a vision [de Buffon] se prolonge vers l'avenir, vers cette Terre de plus en plus froide où l'homme et même toute la Nature vivante lutteront

1. Jacques Delille, 'Chant premier: la lumière et le feu', in *Les Trois règnes de la nature*, vol.1 (Paris, 1808), p.41-76 (45).
2. On theories of neptunism and vulcanism in the later eighteenth century, see Roger, *Buffon* p.510, and Rudwick, *Bursting the limits of time*, p.141.
3. Although much anticipated, the 'Epoques de la nature' were not warmly greeted by critics, largely because the text was perceived as an attack on religious chronology. The Sorbonne affirmed, in November 1779, that Buffon's reading of Scripture and his division of the Creation into seven epochs, in the place of days, were unacceptable. Buffon rather blatantly made no effort to respond to or to placate his accusers. See Roger, *Buffon*, p.555, and Rudwick, *Bursting the limits of time*, p.129.

> pour leur survie. Car à mesure que l'océan continue à baisser, que les végétaux transforment la matière brute en matière combustible, une réserve de chaleur s'accumule. Mais le refroidissement inexorable aura sans doute le dernier mot.
>
> [Buffon's] vision persists into the future, toward a colder and colder Earth where man and even all of living Nature will struggle to survive. For as sea levels continue to drop, as plants transform raw matter into combustible matter, a reserve of heat accumulates. But the inexorable cooling will surely have the last word.[4]

In the 1778 version of the Earth's history, the present was no longer a repetition of the past, but a new and distinct episode, epoch or chapter in the narrative of natural history. It was defined by its temperature, relative to previous epochs, and by which natural activities could occur at this degree of heat. Understanding the present depended upon the ability to read its precise position in this narrative: it was contingent on the chapters that preceded it, and it would in turn give shape to future chapters and epochs of history. However, if reading the natural world and telling its story did not have the same consequences in the 'Epoques' as they did when Buffon began writing the *Histoire naturelle*, the natural historian's task continued to be founded upon the same principle of bringing his work into harmony with the work of nature, through the process of style and through the medium of heat, the energy of natural language. The 'Epoques' may be seen as the culmination of this poetic process because it marked the moment where, for Buffon, written language and created nature finally converged. The heat of Buffon's language of nature, once simply a metaphor for transmitting knowledge of nature, could now be read as operating in the same way as nature itself, and even as bringing to nature the heat it no longer possessed on its own. Both nature and the natural historian produced through heat; now, these two types of heat were meant to be perceived as indistinguishable. Buffon took a step in the 'Epoques' of which he could not have conceived earlier. His language not only made nature legible to the human mind, but allowed nature to be understood as a text with a definite discursive structure. As a result, the objects and events of the world no longer needed to be translated into human language: the two had been made one and the same through heat.

4. Roger, *Buffon*, p.569.

Why did the reader need to be able to read nature through and as a text? By the end of the sixth and penultimate epoch, the warming power of nature to create natural historical phenomena had been dwindling, as it had been ever since fragments of molten rock, the material that had formed the Earth, left the Sun. Yet to draw this conclusion required the ability to visualize changes in nature over time which, on the level of human understanding (limited, not divine, perception), implied reading nature as a language of events sequenced across the pages of a book. Reading nature as a text allowed its activities to be organized and ordered, so that each event might be made sense of with respect to previous ones. The story of the Earth might now be read through degrees of heat: the temperature at a given historical moment indicated how much productive energy was present, determining the kind and quantity of natural activity that could be accomplished at that specific time. Changes in degree of heat affected nature's inventive power, bringing about changes in the Earth's geology, geography and organic systems: first active volcanoes, then cooler and less active mountains, then the formation of large bodies of water, then the appearance of forms of organic life, and eventually the presence of human beings. The 'Epoques' were fashioned as a guide for reading nature over the course of its history. The activity and thermal energy of each epoch was vividly reproduced in the reader's mind through the sequence of chapters in the book, as through the style of Buffon's language.

Yet the idea behind the 'Epoques' was not just to read a gripping story; its language had fundamental epistemological and moral goals. Through his writing, Buffon reproduced the energy of natural production, thus imbuing readers with a discursive energy that they might learn to intelligently redeploy in the world around them. If they could understand how to translate language back into reality, the vivacity of narrative back into physical activity, Buffon's dream was that people could delay nature's death by producing their own heat. They could figure out how to make nature work to their advantage and how to surpass it, by becoming not only the users of heat, but its producers and regulators. Buffon wrote:

> C'est de la différence de température que dépend la plus ou moins grande énergie de la Nature; l'accroissement, le développement et la production même de tous les êtres organisés ne sont que des effets particuliers de cette cause générale: ainsi l'homme en la modifiant, peut en même temps détruire ce qui lui nuit et faire éclore tout ce qui lui convient.

> It is on the difference in temperature that the more or less great energy of Nature depends; the growth, the development and even the production of all organized beings are nothing but the specific effects of this general cause: thus man, by modifying it, can at the same time destroy what is detrimental to him and make flourish everything that suits him.[5]

How could people learn to adjust the energy of nature and to control the temperature of a world so much bigger than themselves? What might turn them from passive readers of history to active transformers of its course? An answer can be found in examining how Buffon's language, especially in the last chapter of the 'Epoques', used heat to make the transition from composer of history to author of the world's future.

Buffon's theorization of heat

The presence and activity of the Earth's heat had become quite significant in Buffon's work by the 1770s. He was inspired in part by the work of his colleague, Dortous de Mairan, at the Royal Academy of Sciences of Paris. Mairan's 1765 'Nouvelles recherches sur la cause générale du chaud en été et du froid en hiver' ('New research on the general cause of heat in summer and cold in winter') argued that the Earth did not receive nearly enough warmth from the Sun to sustain its current temperature, and that therefore it had to have its own, inner fire.[6] Buffon wrote of Mairan's findings that

> l'on doit reconnoître deux sortes de chaleur, l'une lumineuse, dont le soleil est le foyer immense, et l'autre obscure, dont le grand réservoir est le globe terrestre. Notre corps, comme faisant partie du globe, participe à cette chaleur obscure; et c'est par cette raison qu'étant obscure par elle-même, c'est-à-dire, sans lumière, elle est encore obscure pour nous, parce que nous ne nous en apercevons par aucun de nos sens. Il en est de cette chaleur du globe comme de son mouvement, nous y sommes soumis, nous y participons sans le sentir et sans nous en douter. De-là il est arrivé que les Physiciens ont porté

5. Buffon, 'Des époques de la nature', p.245-46.
6. See Jean-Jacques Dortous de Mairan, 'Nouvelles recherches sur la cause générale du chaud en été et du froid en hiver', in *Histoire de l'Académie royale des sciences, année MDCCLXV, avec les mémoires de mathématique & de physique pour la même année* (Paris, Imprimerie royale, 1768), p.143-266. Roger discusses this work in *Buffon*, p.508-509.

> d'abord toutes leurs vues, toutes leurs recherches sur la chaleur du soleil, sans soupçonner qu'elle ne faisoit qu'une très-petite partie de celle que nous éprouvons réellement.

> [W]e must recognize two sorts of heat, the first luminous, of which the Sun is the immense hearth, and the other obscure, the great reserve of which is the terrestrial globe. Our bodies, as they are part of the globe, participate in this obscure heat; and it is for this reason that, being obscure in itself, that is, without light, it remains obscure for us because we cannot perceive it through any of our senses. The same applies to the movement [of this obscure heat]: we are subject to it, we participate in it without sensing it and without doubting it. From here it has come to pass that Physicists first directed all their attention, all their research, toward the heat of the Sun, without suspecting that it only comprises a very small part of that [heat] which we effectively feel.[7]

Soon after reading Mairan's work, Buffon began experiments on globes of iron and then on globes of the other minerals of which the Earth was composed and which could be found at different depths and in different strata relative to the surface of the planet.[8] He observed how long it took for each type of mineral to heat up, then to cool and to become tolerable to human touch, in other words, mimicking the relatively warm, temperate state of the present-day Earth. This not only showed how internal heat moved and behaved within each type of mineral, giving it a specific character and role, but also allowed Buffon to sequence and construct a narrative of which substances cooled first (that is, further back in time) in the history of the world. In general, these could be found deeper within the layers of the Earth's crust, and from here he deduced and calculated the order and duration of the solidification of each mineral with respect to the others. The amount of time it took each mineral to cool created natural starting and ending points for the chapters of the 'Epoques'. Buffon represented the activities possible in each of these chapters or historical durations with the lively heat of his prose, reenacting the

7. Buffon, 'De la lumière, de la chaleur et du feu', in *Histoire naturelle, générale et particulière: supplément*, vol.1 (Paris, Imprimerie royale, 1774), p.1-78 (32-33).
8. On Buffon's experiments, see Roger, *Buffon*, p.517-18, and Lucien Leclaire, 'L'*Histoire naturelle des minéraux* ou Buffon géologue universaliste', in *Buffon 88*, ed. J.-C. Beaune *et al.*, p.355-65. Buffon also recounted the results of these experiments in 'Des verres primitifs', in his *Histoire naturelle des minéraux*, vol.1, p.17-29.

succession of events that would have been possible at a certain moment in time and temperature.

Buffon published the results of these experiments, as well as reflections upon them, in 1774 and 1775, in the first two volumes of the *Supplément* to the *Histoire naturelle.* In the introductory essay from the first volume (1774) of the *Supplément*, entitled 'De la lumière, de la chaleur et du feu', he explained the operation of nature through the movement of matter, recalling the mechanical explanation of fire given by D'Alembert in the *Encyclopédie* article 'Chaleur' (see above, Chapter 2), as well as a discussion of the impulsive motion of the comet in the previous chapter. In particular, mirroring what he had already written about the movement and development of the solar system (and echoing his translation of Hales' *Vegetable staticks*), Buffon wrote of two 'forces primitives, celle qui cause la pesanteur, et celle qui produit la chaleur' ('primitive forces, that which causes gravity, and that which produces heat'). The first of these forces he named 'attraction', or gravitation, and the second 'impulsion', as he had done in 1749 when describing the movement of the comet – except that now the concept of heat was explicitly added to these forces. While attraction pulled particles of matter toward one another, impulsion worked 'au moyen du ressort' ('like a spring'), pushing particles away from one another, an action which produced heat.[9] Buffon described impulsion as 'une force expansive' ('an expansive force') because it pushed particles 'du centre à la circonférence' ('from the center to the circumference'). Matter subject to gravitation alone, whose particles moved 'au contraire de la circonférence au centre' ('by contrast from the circumference to the center'), was described as 'brute'.[10] In this way, impulsion was an anti-gravitational force: 'la force expansive est la même que la force attractive devenue négative' ('the expansive force is the same as the attractive force, rendered negative').[11] These two primitive forces of nature were presented like the two sides of a same coin: 'ce sont pour la Nature deux instrumens de même espèce, ou plutôt ce n'est que le même instrument qu'elle manie dans deux sens opposés' ('they are for Nature two instruments of the same type, or rather they are but the same instrument that she [Nature] manipulates in two different directions').[12]

9. Buffon, 'De la lumière, de la chaleur et du feu', p.1-2.
10. Buffon, 'De la lumière, de la chaleur et du feu', p.4-5.
11. Buffon, 'De la lumière, de la chaleur et du feu', p.18.
12. Buffon, 'De la lumière, de la chaleur et du feu', p.15.

Heat, according to this philosophy, was a change in the state of matter in a body provoked through the communication of movement of molecules endowed with impulsive force.[13] The state of matter was modified when the activity of impulsion compelled molecules, pulled together by gravitation, to push off of one another in a spring-like manner, and to enter new bodies, in turn making the particles of these bodies move faster and creating heat within them. Buffon concluded, moreover, that the entire universe could be understood as different forms and degrees of heat, controlled by the same law of attraction and impulsion: 'la lumière, [...] la chaleur et le feu ne sont que des manières d'être de la matière commune; qu'il n'existe en un mot qu'une seule force et une seule matière toujours prête à s'attirer ou à se repousser' ('light, [...] heat and fire are nothing but states of communal matter; that there exists, in a word, only one single force and one single matter that is always ready to attract or to repulse').[14]

Impulsion explained not only the solidification of minerals and hence the formation of the Earth's geology and geography, but also the genesis of organic matter and living organisms. The energy of impulsion transformed raw or 'brute matter' ('la matière brute') into 'live matter' ('la matière vive'). Live matter was characteristic of 'tous les êtres qui vivent ou végètent, mais encore toutes les molécules organiques vivantes, dispersées et répandues dans les détrimens ou résidus des corps organisés; je comprends encore dans la matière vive, celle de la lumière, du feu, de la chaleur, en un mot toute matière qui nous paroît active par elle-même.'[15] Impulsion endowed bodies with heat, distinguishing them from the stable background of attractive force, and allowing them to develop and to change. '[P]ar le moyen de la chaleur,' Buffon continued, 'l'impulsion pénètre dans les corps

13. Buffon, 'De la lumière, de la chaleur et du feu', p.31.
14. Buffon, 'De la lumière, de la chaleur et du feu', p.18. On Buffon's construction of the universe through the equilibration of opposing forces, see the article 'Feu' in the *Dictionnaire de chimie* of Buffon's chemist colleague Pierre-Joseph Macquer (2 vols, Paris, Lacombe, 1766), vol.1, p.498-506; Roger, *Buffon*, p.510-12; and Jean Svagelski, *L'Idée de compensation en France, 1750-1850* (Paris, 1981), p.191-94.
15. '[A]ll beings that live or vegetate, and furthermore all organic living molecules dispersed and spread out among the detriments or residues of ordered bodies; by live matter, I also mean that of light, of fire, of heat, and in one word all matter that seems to us to be independently active.' Buffon, 'De la lumière, de la chaleur et du feu', p.4. For a summary of Buffon's theory of attraction and expansion, and how it applied to life (*matière vive/matière brute*) and living molecules, see Thomas S. Hall, *Ideas of life and matter: studies in the history of general physiology, 600 BC-1900 AD*, vol.2 (Chicago, IL, 1969), p.5-17.

organisés, [et] c'est par la chaleur qu'ils se forment, croissent et se développent' ('[B]y means of heat [...] impulsion penetrates organized bodies, [and] it is through heat that they form, grow and develop').[16] He imagined the world as a continuous flow of energy through bodies of matter. There was no such thing as being completely dead or completely inorganic – all substances contained within them some particles of ether or heat. Thus Buffon described minerals, especially those composed of 'la terre végétale' – humus, the decomposed trace of formerly more 'organized' beings – as alive, albeit to a lesser degree than plants and animals:

> Les productions de la Nature organisée qui, dans l'état de vie et de végétation, représentent sa force et font l'ornement de la terre, sont encore, après la mort, ce qu'il y a de plus noble dans la Nature brute: les détrimens des animaux et des végétaux conservent des molécules organiques actives, qui communiquent à cette matière passive, les premiers traits de l'organisation.

> The productions of organized Nature which, in the states of life and vegetation, represent her force and decorate the earth, are still, after death, that which is the most noble of brute Nature: the detriments of animals and plants conserve active molecules which communicate the initial qualities of organization to this passive matter.[17]

Within the earth, minerals formed through the Newtonian action–reaction dynamic: inorganic particles pulled together while the elements of 'la terre végétale' whispered the heat and movement of life, aspiring toward more organization, toward a vegetal and eventually a vital state: 'la matière ductile, pénétrée et travaillée dans tous ses points, et par conséquent dans les trois dimensions à la fois, prend la forme d'un germe organisé, qui bientôt deviendra vivant ou végétant par la continuité de son développement et de son extension proportionnelle en longueur, largeur et profondeur.'[18] Minerals, and especially fossils, still retained degrees of the heat of this past life. They

16. Buffon, 'De la lumière, de la chaleur et du feu', p.4.
17. Buffon, *Histoire naturelle des minéraux*, vol.1, p.4-5. On humus, see 'De la terre végétale', in Buffon, *Histoire naturelle des minéraux*, vol.1, p.384-426.
18. '[D]uctile matter, infused and worked [with life] in all of its parts, and thus in all three dimensions at once, takes on the form of an organized seed that will soon become alive or vegetative through the continuity of its development and its proportional growth in length, height and depth.' Buffon, *Histoire naturelle des minéraux*, vol.1, p.5.

were a representation or 'figuration' of life; not yet organized enough to pass into the realm of plants, the vital forces remaining within the fossil gave to its minerals a superficial, 'exterior appearance' ('forme extérieure') which made them resemble life: 'cette matière n'étant travaillée qu'à la surface ne pourra prendre d'autre forme que celle d'un minéral figuré' ('this matter, having only been worked on its surface, can take no other form than that of a figured stone').[19] Buffon would have read a fossil of an ancient fern, for example, as still able to represent living form, even though it had lost much of the heat of life itself. Most of the fossil was rock, but the vitality of the fern had not completely departed: some living molecules remained, keeping at least the external, visible shape of the plant in place and reminding the natural historian of the former heat and potential of nature.

Growth, development, change and the vital flow of energy across the chains of being and time were key to the function of heat in the 'Epoques de la nature'. However, the apparently simple language of 'one single force and one single matter' presented in 'De la lumière, de la chaleur et du feu' did not mean that attraction and impulsion balanced each other out, or that the movement of heat in the 'Epoques' was progressing toward a stable economy or equilibrium. Quite the opposite: Buffon's nature was dying from loss of heat, its energy finite, its end inevitable. However, the end could be postponed as people learned to imitate and appropriate the force of nature's laws. Impulsion, through heat, was the force which wrote the events of nature's history.[20] These were no longer the continuous, circular, predictable (because repetitive) events of 1749 epitomized in the ebb and flow of tides or the growth and erosion of continents and in the theories of heat, energy and movement to which Buffon had been exposed earlier in his career. Events in the 'Epoques' were singular moments that were contingent upon what came before and that modified what would come. From a circular, contained system, Buffon

19. Buffon, *Histoire naturelle des minéraux*, vol.1, p.5.
20. Michel Delon describes Enlightenment history as written through the energy created by the opposition of forces (attraction and inertia, heat and coldness, violence and refinement in civilizations), which 'désigne un équilibre qui n'en finit pas de se chercher, un mouvement par approximations successives. L'énergie se contente d'assurer [...] la mutabilité des choses; elle devient [...] impulsion d'un devenir historique' ('designates an equilibrium that never ceases to try to find itself, a movement by successive approximations. Energy is content to guarantee [...] the mutability of things; it becomes [...] the impulsion of a historical becoming'). See Delon, *L'Idée d'énergie en France*, p.210-11.

moved to a more open-ended one whose operation required a new inventive force. Understanding the physical process through which the Earth's history had been generated and learning to read and interpret its events, written in the natural language of heat, were the first steps in showing people how to assume control over natural productions and how to take over the creation of history from a world that was slowly losing its impulsive force. This was accomplished through the rhetorical impulsion of Buffon's language.

The natural history of human beings: a story of inventing the temperate

At the opening of the final chapter of the 'Epoques de la nature', the world had finally become habitable to humans. Nature had almost overcome its brutal and intemperate past. The matter which comprised nature had less impulsion, and thus produced less heat with which to provoke violent, revolutionary changes (volcanoes, earthquakes, floods) to the Earth. Nature had become calmer through a gradual, slow loss of heat: as Buffon wrote, 'ce refroidissement si lent' ('this very slow cooling') had occurred over hundreds of thousands of years 'd'une manière insensible' ('in an imperceptible manner').[21] Over time, the inventive and transformative power of fire (one of the states of matter that was produced by the impulsive force of heat) had been subdued. 'Il a fallu six cents siècles à la Nature', Buffon explained, 'pour construire ses grands ouvrages, pour attiédir la Terre, pour en façonner la surface et arriver à un état tranquille' ('Nature required 600 centuries to construct its great works, to cool the Earth, to shape its surface, and to come to a tranquil state').[22]

Because it had lost so much of its force and violence over time, in the final chapter of the 'Epoques' the nature of the present era no longer held the ultimate authority to write the events of history and bring about changes to the Earth. Natural history was now amenable to coauthorship, for Buffon implied that the laws once controlled by nature were now able to be modified by the heat-generating activities of human thought and action. However, in the 'Septième époque' the problem remained that people had not been, historically, in a state to understand and to alter the world in a fruitful, dynamic manner.

21. Buffon, 'Des époques de la nature', p.242.
22. Buffon, 'Des époques de la nature', p.238.

'[C]ombien [de temps]', asked Buffon, 'n'en faudra-t-il pas pour que les hommes arrivent au même point [que la nature] et cessent de s'inquiéter, de s'agiter, et de s'entre-détruire?' ('[H]ow much [time] will be necessary until men come to the same point [as nature] and stop worrying, provoking and destroying one another?')[23] In order to surpass nature, or at least to partake in its work, people had first to learn to think in its language and to reflect upon their own.

The image at the beginning of the 'Septième époque' of the first people, living in the Enlightenment's much imagined state of nature before the invention of society, is striking. In a critical tone, Buffon described these people, who appeared in the narrative right at the cusp of nature's transition to a new, less fiery and thus more peaceful, epoch. He depicted them as helpless before 'des mouvemens convulsifs de la Terre' ('the convulsive movements of the Earth'), 'les innondations' ('floods') and 'le feu des volcans' ('volcanic fire').[24] In this still volatile world, the first people were simply the products of nature and were dominated by its force. They were 'tremblans sur une terre qui trembloit sous leurs pieds, nus d'esprit et de corps, exposés aux injures de tous les élémens, victimes de la fureur des animaux féroces, dont ils ne pouvoient éviter de devenir la proie; tous également pénétrés du sentiment commun d'une terreur funeste, tous également pressés par la nécessité'.[25] Buffon narrated the story of how, slowly, these people came together to protect themselves against the world. They began to act through heat, writing with it the history of their presence on Earth. From very early on, they were able to imitate the power of nature, producing heat through the force of impulsion in the same manner as nature. For example, without realizing the implications of the activity of hitting stones together, people created a shock which gave the molecules in the stones more spring-like movement, causing them to push off one another and to move outwards toward other bodies of matter likely to be affected by heat such as wood. Replicating the natural force of impulsion, this movement brought about fire, and eventually, 'avec le secours de ce

23. Buffon, 'Des époques de la nature', p.238.
24. Buffon, 'Des époques de la nature', p.225.
25. '[T]rembling upon an earth that trembled beneath their feet, naked in mind and body, exposed to the offenses of all the elements, victims of the furor of ferocious animals of which they could not help but become the prey; all equally filled with the shared sentiment of a gruesome terror, all equally pressured by necessity.' Buffon, 'Des époques de la nature', p.225.

puissant element, il [l'homme] a nettoyé, assaini, purifié les terreins qu'il vouloit habiter' ('with the help of this powerful element, man cleaned, sanitized, purified the lands he wanted to inhabit').[26] People cleared terrain and reordered the land through deforestation, burning, agriculture, herding and building, all forms of heat. They gradually learned to imitate and reproduce the energy of the world around them and slowly generated a collective power to leave the trace of their presence and activity upon the surface of the world, as upon the pages of the *Histoire naturelle.* Through the expansive force of their activity, different groups of people soon collided and confronted each other for the use of resources which they had to learn how to divide, share or dominate. From here, Buffon attributed the development of societies, nations, laws and order ('la police'), and 'c'est de ce moment que la Terre est devenue le domaine de l'homme' ('it is at this moment that the Earth became the domain of mankind').[27] Yet, for all that people had accomplished in harnessing the power of nature throughout the history of natural, civil and political societies, Buffon insisted that they were not often aware of the work they did, and could, accomplish in imitating and paralleling nature. Only in the present moment, at the end of the 'Septième époque', did he deem people finally capable of perceiving nature's history and laws on a more general scale, and thus of grasping the true consequence of their place, activity and authority on Earth. It was only at this point that they could actively comprehend the process through which nature might be modified, controlled and transformed to their benefit.

According to Buffon there was another moment, earlier in the history of mankind, when a great civilization thrived that seemed to understand its relationship to the natural climate better than any of its predecessors or successors. It was located 'dans le centre du continent de l'Asie, depuis le 40e degré de latitude jusqu'au 55e' ('in the center of the Asian continent, from the 40th degree of latitude to the 55th'), and was part of 'la Sibérie méridionale et de la Tartarie' ('southern Siberia and Tartary').[28] Buffon described this nation as a secular Eden, a paradise in historical time, in which all the elements of a perfect,

26. Buffon, 'Des époques de la nature', p.226.
27. Buffon, 'Des époques de la nature', p.226.
28. Buffon, 'Des époques de la nature', p.228-29. On a modern-day map, according to Roger, this country would be between Tajikistan and Russia, 'entre le plateau de Pamir, au Sud-Ouest, et le lac Baikal, au Nord-Est'. See Buffon, *Les Epoques de la nature*, ed. J. Roger, p.306 (n.6).

temperate climate had happily converged.[29] This 'contrée priviligiée, à l'abri des inondations, [était] éloignée des volcans, plus élevée, et par conséquent plus anciennement tempérée que les autres' ('privileged country, protected from floods, [was] far away from volcanoes, more elevated, and consequently had a temperate climate well before other countries').[30] At a moment when, according to the historical trajectory of the 'Epoques', the Earth was still slightly warmer near the poles and in its northern regions, the climate of this country was neither too warm nor too cold and was perfectly suited not only to overcoming the terrors of nature, but to producing intelligence, peace and science. Because they were not threatened by the violent energy of which nature was capable, its inhabitants were able to take time to understand nature's order and power and to develop a similar intellectual power, harmonizing their order of knowledge with the order of the world. 'Puisqu'il est devenu très-savant' ('Because it became so learned'), this people was 'très-heureux' ('very happy'), and 'a joui pendant plusieurs siècles de la paix, du repos, du loisir nécessaire à cette culture de l'esprit de laquelle dépend le fruit de toutes les autres cultures' ('over several centuries enjoyed peace, calm and the leisure necessary to that culture of the mind upon which depends the fruit of all other cultivation').[31] Following the agricultural metaphor in this quotation, to best cultivate the earth and thus to make nature more productive, fashioning it to serve human needs, people had to first understand how to culture or cultivate, to order and enrich, their perception. They had to first be able to visualize and speak of the world as their garden, before transforming the poetic energy of thought into the physical, impulsive, expansive energy of natural creation. The real geographic location of this north Asian civilization was as important as its symbolic value. It was where Buffon envisioned the site of earthly paradise, defined not through God or human sin but through climatic conditions at a particular time and place in the historical trajectory of the Earth's heat. This paradise existed because of a particular temperature perfectly suited to material and intellectual production.

29. Buffon was not the only thinker at the time to imagine a historical Eden, or to trace the echoes of prelapsarian intelligence as it dissipated through ancient societies and time. Roger attributes Buffon's placement of the first great civilization to his reading of the astronomer Jean-Sylvain Bailly's *Histoire de l'astronomie ancienne* (1775) and *Lettres sur l'origine des sciences et sur celle des peuples de l'Asie* (1777). See Roger, *Buffon*, p.551.

30. Buffon, 'Des époques de la nature', p.228.

31. Buffon, 'Des époques de la nature', p.230.

It was a moment in which knowledge of nature was created not through a direct communication between God and man, but through humans learning to read the book of nature through experience and interaction with the world.

What happened to this ideal civilization? Why were its knowledge and happiness not passed down through the generations? Why did only 'débris et quelques résultats qu'on pouvoit retenir de mémoire' ('debris and a few traces that could be remembered') remain?[32] Using the example of astronomy (in this science he saw some of the greatest achievements of this lost civilization), Buffon asked how today it was 'encore bien imparfaite après deux mille ans de nouvelle culture et même jusqu'à nos derniers siècles' ('still quite imperfect, after 2000 years of culture and even in the last centuries [of our civilization]').[33] The earthly paradise fell because nature continued its gradual cooling, and the parts of the globe closer to the poles experienced drops in temperature. With this decline in temperature came a decline in scientific activity. The activities of humans were still, at this historical moment, largely controlled by the power of nature; much of the productive force that created the culture of both land and mind in this country was the result of favorable climatic conditions. This was a paradise authored only in small part by humans, for they were not yet aware of the extent of their authorial capabilities.

Moreover, Buffon envisioned how the countries to the north of this civilization, whose cooling climates in turn cooled the mental capability and dexterity of their inhabitants, would have become greedy and violent, unable to perceive the general importance and potential of the paradise just to the south: 'La perte des sciences, cette première plaie faite à l'humanité par la hache de la barbarie [aura eu lieu] quand les terres situées au nord de cette heureuse contrée ont été trop refroidies, [et] les hommes qui les habitoient, encore ignorans, farouches et barbares auront reflué vers cette même contrée riche, abondante et cultivée par les arts.'[34] These invasions would have succeeded in extinguishing 'non-seulement les germes, mais même la

32. Buffon, 'Des époques de la nature', p.233.
33. Buffon, 'Des époques de la nature', p.233.
34. 'The loss of sciences, that initial wound done to humanity by the axe of barbarians [would have happened] when the lands situated to the north of this happy country cooled down too much, [and] the men who inhabited them, still ignorant, wild, and barbaric, would have surged back into this rich, abundant, and artistically cultivated country.' Buffon, 'Des époques de la nature', p.233-34.

mémoire de toute science' ('not only the seeds, but indeed the memory of all science').[35] They left only 'les rameaux stériles ou dégénérés des nobles branches de cette ancienne souche' ('the sterile or degenerated branches of the noble boughs of this ancient lineage'). From this corrupted ancestry 'l'homme d'alors replongé dans les ténèbres de l'ignorance a pour ainsi dire cessé d'être homme' ('man, now plunged back into the darkness of ignorance, ceased as it were to be man').[36]

When people lost their nature and their identity, they were no longer able to think or create with style. Recalling the 'Discours sur le style', the process of building a beautiful and functional textual world in nature's stead was not about formal imitation or superficial ornament and decoration. Buffon wrote that 'le style, c'est l'homme': without deliberate awareness of its productive power, without actively understanding the mechanisms of past prosperity, humankind would lose its fundamental meaning and essence. With the fall of Buffon's great imagined civilization, style also disappeared. People partially regressed to the state of nature described at the beginning of the 'Septième époque'. This state was based upon emotion (namely fear), a direct, unmediated and unconscious relationship with the natural world. The first people were not aware of their ability to understand how the world worked and to bring it under their control. They were afraid of the earthquakes, volcanoes and floods threatening to destroy them. The processes of admiration, experience and reason, of slowly becoming acquainted with nature, of creating a vision of the world through language, was latent. Without this reflexive capability, Buffon wrote, people continued and would continue to see and to act through the lens of terror: '[C]es hommes profondément affectés des calamités de leur premier état, et ayant encore sous leurs yeux les ravages des inondations, les incendies des volcans, les gouffres ouverts par les secousses de la Terre, ont conservé un souvenir durable et presque éternel de ces malheurs du monde.'[37] Terror engendered veiled and obscure ways of explaining the world, narratives whose power derived from a profound persuasive weight in the form of faith and myth.[38] For

35. Buffon, 'Des époques de la nature', p.233-34.
36. Buffon, 'Des époques de la nature', p.234-35.
37. '[T]hese men, profoundly affected by the calamities of their initial state, and still faced with the ravages of floods, the fires of volcanoes, the caverns opened by the shakings of the Earth, held on to an enduring and almost eternal memory of these evils of the world.' Buffon, 'Des époques de la nature', p.227.
38. Buffon may have taken this idea of how early people read the activities of nature from a manuscript copy of Nicolas-Antoine Boulanger's *Anecdotes de la nature.*

instance, the vast, mysterious ocean could be read as the remains of the great biblical flood and earthquakes could be understood as God's way of continually punishing people for their sins.[39] According to Buffon, faith in these explanations blinded people, as they were unable to see that such fabulous and miraculous stories did not constitute reasoned, measured knowledge but rather were the inventions of their own emotions. They could be read as intellectual markers of how nature was understood in the past.[40] He referred to these narratives as 'opinion', not 'science': 'l'opinion de l'existence réelle d'un Etre mal-faisant, la crainte et la superstition qui en sont le premier produit' ('the opinion of the real existence of an evil Being, the fear and the superstition that directly resulted from it'), were passed down through the centuries of human history. '[T]ous ces sentimens fondés sur la terreur', he continued, 'se sont dès-lors emparés à jamais du cœur et de l'esprit de l'homme' ('[A]ll of these sentiments founded on terror seized thenceforth and forever the heart and the mind of man').[41]

In the present day, Buffon thought that humankind was only just beginning to find its way out of this paralyzing fear and to learn to comprehend nature through the narrative of the steady collection and association of observations over time, as articulated in the first volume of the *Histoire naturelle*. '[A] peine est-il encore aujourd'hui rassuré par l'expérience des temps, par le calme qui a succédé à ces siècles d'orages,

---

This unpublished work was believed to be lost after Boulanger's death in 1759. Buffon was accused of having plagiarized from a circulated copy to write the 'Epoques'. See Roger, *Buffon*, p.551. On Boulanger, see Jacques Roger, 'Un manuscrit inédit perdu et retrouvé: les *Anecdotes de la nature* de Nicolas-Antoine Boulanger', *Revue des sciences humaines* 71 (July-September 1953), p.231-54, and Paul Sadrin, *Nicolas-Antoine Boulanger (1722-1759), ou Avant nous le déluge*, *SVEC* 240 (1986), p.205-21.

39. On Enlightenment interpretations of the biblical flood, see Nicolas-Antoine Boulanger, 'Déluge', in Diderot and D'Alembert, *Encyclopédie*, vol.4, p.795-803; Cohn, *Noah's flood*; and Maria Susana Seguin, *Science et religion dans la pensée française du XVIIIe siècle: le mythe du déluge universel* (Paris, 2001).

40. This argument, from the principle of accommodation, was famously made by Galileo, who wrote that the Bible could not be used to dismiss his conclusions about the physical state of the universe because it had been written in an earlier era, when people possessed less knowledge. The writers of the Bible had to adjust their language to their public's intellectual capacity, making an understanding of the world accessible to them. See Galileo's 'Letter to the Grand Duchess Christina', in *Discoveries and opinions of Galileo*, ed. and translated by Stillman Drake (New York, 1957), p.173-216.

41. Buffon, 'Des époques de la nature', p.227.

enfin par la connoissance des effets et des opérations de la Nature' ('[Man] today has just barely been reassured by the experience of time, by the calm that followed these stormy centuries, and finally by the knowledge of the phenomena and operations of Nature').[42] He implied that the language and style of the *Histoire naturelle* could become a guide to escaping from centuries of irrationality and violence, showing how to reread history and thus how to assimilate the ability to create it.[43] Buffon's readers had to learn to transform what had traditionally been perceived with terror (earthquakes, comets, oceans, large animals) into anecdotes, useful lessons, that could help them in the face of a cooling globe. He modeled this kind of work from the very beginning of the 'Septième époque', showing how people in the state of nature, although they were afraid of fire, yet managed to use it to tame the world around them. Although they did not realize it, these early people had begun to appropriate nature's power. Anecdotes such as this became monuments in the text, rhetorical markers in a persuasive and hopeful story not of the inevitable end of the world, but of people's increasing ability to come to nature's aid. The story was designed to appeal to the reader's sensibility, not through faith in nature and God, but through both the affective, persuasive power of Buffon's language and its rigorously logical and methodological structure. Buffon moved his readers' emotions in order that they consider the Earth's past in a coherent and systematic manner and understand how to act rationally upon its present.

It was never enough, according to Buffon, to have knowledge, or even to have the potential to know. Action, or the material realization of knowledge, was and always would be, crucial. 'Plus il [l'homme] a su, plus il a pu,' Buffon remarked, 'mais aussi moins il a fait, moins il a su' ('The more he [man] knew, the more he was able, but also the less he did, the less he knew').[44] Doing, embodied by the verb 'faire', was an essential part of knowing. The less one did, the less one knew, and hence, it may be assumed, the less potential – the verb 'pouvoir' – one had either to know or to do. Buffon emphasized the moments in history when people had successfully monumentalized their knowledge, leaving its impression upon the surface of the Earth, and thus creating a narrative of accomplishments that might be read

42. Buffon, 'Des époques de la nature', p.227.
43. Wilda Anderson explains that, in the 'Epoques', '[h]istory is both the story of the past and the training-ground for future action' ('Error in Buffon', p.701).
44. Buffon, 'Des époques de la nature', p.228.

not as subservient to the works of nature, but as nearly equal to them. In these moments, 'la puissance de l'homme s'est réunie à celle de la Nature' ('the power of man joined with that of Nature'):

> [L]es trésors de sa fécondité jusqu'alors étoient enfouis, l'homme les a mis au grand jour; [...] par-tout, lorsqu'il s'est conduit avec sagesse, il a suivi les leçons de la Nature, profité de ses exemples, employé ses moyens, et choisi dans son immensité tous les objets qui pouvoient lui servir ou lui plaire. Par son intelligence, les animaux ont été apprivoisés, subjugués, domptés, réduits à lui obéir à jamais; par ses travaux, les marais ont été desséchés, les fleuves contenus, leurs cataractes effacées, les forêts éclaircies, les landes cultivées; [...] enfin *la face entière de la Terre porte aujourd'hui l'empreinte de la puissance de l'homme*, laquelle, quoique subordonnée à celle de la Nature, souvent a fait plus qu'elle, ou du moins l'a si merveilleusement secondée, que c'est à l'aide de nos mains qu'elle s'est développée dans toute son étendue, et qu'elle est arrivée par degrés au point de perfection et de magnificence où nous la voyons aujourd'hui.

> [T]he treasures of his productivity had been buried until then, man brought them to light; [...] everywhere, when he acted with wisdom, he followed the lessons of Nature, profited from its examples, used its means, and selected from within its immensity all the objects that could help or please him. Through his intelligence, animals were tamed, subjugated, controlled, made to forever obey him; through his works, swamps were dried, rivers contained, their cataracts eliminated, forests cleared, heaths cultivated; [...] ultimately *the entire face of the Earth carries today the traces of the power of man*, which, although subordinate to the power of Nature, had so often done more than it, or at least had so marvelously assisted Nature, that it is with the help of our hands that Nature has developed in all its extent, and that it has come by degrees to the point of perfection and magnificence in which we see it today.[45]

Buffon wrote with an inspirational tone, revealing to his readers the feats people had already achieved, arguing that it was no small thing to have altered and rewritten the pages of the book of nature and to have learned to coexist beside nature as an author of terrestrial phenomena. He made his readers aware of their past, so as to better understand and to ameliorate the future. Yet, beyond recognizing the feats already accomplished, he wanted to show them the process and

45. Buffon, 'Des époques de la nature', p.236-37 (my emphasis).

the method of these accomplishments. The instances where humans tamed nature were the results of repeated observation, experience, trial and error, the constant process of '[mettre le savoir] au grand jour' ('[bringing knowledge] to light'). Writing about how people produced new kinds of fruit trees through grafting, Buffon remarked:

> [C]ombien de fois n'a-t-il pas fallu que l'homme ait tenté la Nature pour en obtenir ces espèces excellentes? combien de milliers de germes n'a-t-il pas été obligé de confier à la terre pour qu'elle les ait enfin produits? Ce n'est qu'en semant, élevant, cultivant et mettant à fruit un nombre presque infini de végétaux de la même espèce, qu'il a pu reconnoître quelques individus portant des fruits plus doux et meilleurs que les autres.

> [H]ow many times did man have to experiment with Nature in order to obtain these excellent species? How many thousands of seeds was he not obliged to entrust to the earth so that it would finally germinate them? It is only by planting, growing, cultivating and bringing to fruition an almost infinite number of plants of the same species that man was able to recognize a few individuals that had sweeter and better fruits than the others.[46]

Buffon's language insinuated that people had the ability to learn not only to understand the operation of the world and the language of its laws, but to render these laws more energetic and more harmonious with the structure of their thought and their ideas for the future. What, his work implied, would be the consequences of putting some of the authoritative power of nature into the hands of human beings? What did it mean to be not only an observer or reader, but an author, of the world?

Writing the future with heat

Perceiving the world through the lens of a text with style allowed for the identification, ordering and interpretation of changes undergone by nature in the past. In the 'Epoques', the history of nature was arranged and given continuity in terms of heat. The gradual dissipation of heat, starting from the Earth's first moments as a planet in its own right, became a measuring tool which permitted the placement of the chapters of history and the order and tempo of their movement. Heat

46. Buffon, 'Des époques de la nature', p.251.

was the metaphor through which Buffon was enabled to write about nature's actions, productions and changes over time. It was at this moment of passage between the text and the real world that heat and its effects also became real.

Following the thesis of the 'Epoques', people indeed would have to go beyond nature if they were to keep the planet at a warm temperature suitable to both physical cultivation and intellectual culture in the centuries to come. Buffon gave examples from history to convince people of the power they may not have known they possessed. Following a simple, deterministic logic, 'toute la Terre doit être plus froide aujourd'hui qu'elle ne l'étoit il y a deux mille ans; or la tradition semble nous prouver le contraire' ('the whole Earth should be colder today than it was 2000 years ago; however, tradition seems to prove the contrary'):

> Les Gaules et la Germanie nourrissoient des élans, des loups-cerviers, des ours et d'autres animaux qui se sont retirés depuis dans les pays septentrionaux; cette progression est bien différente de celle que vous leur supposez du Nord au Midi. D'ailleurs l'histoire nous apprend que tous les ans la rivière de Seine étoit ordinairement glacée pendant une partie de l'hiver; ces faits ne paroissent-ils pas être directement opposés au prétendu refroidissement successif du globe? Ils le seroient, je l'avoue, si la France et l'Allemagne d'aujourd'hui étoient semblables à la Gaule et à la Germanie; si l'on n'eût pas abattu les forêts, desséché les marais, contenu les torrens, dirigé les fleuves et défriché toutes les terres trop couvertes et surchargées des débris même de leurs productions.

> The Gauls and the Germans fed moose, lynx, bears and other animals who then retreated to northern countries; the movement is quite different from the one you would expect them to make, from north to south. Furthermore, history tells us that every year the Seine was typically frozen during part of the winter; do not these facts appear in direct opposition to the claimed gradual cooling of the globe? They would be, if modern France and Germany resembled Gaul and Germania; if people had not felled the forests, dried up the swamps, contained streams, redirected rivers and cleared the lands that were covered and overwhelmed with the debris of their own productions.[47]

Man, moreover, was capable not only of making heat, but of making it to the degree 'qui lui convient' ('that suits him'), becoming the 'maître

47. Buffon, 'Des époques de la nature', p.240-41.

de l'élément du feu, qu'il peut augmenter et propager à son gré' ('master of the element of fire, that he can augment and propagate at his will').[48] As the initial impulsion of the comet and its fire slowly dissipated, nature grew ever colder. Cold was not just the absence of heat but the absence of doing, of the impulsive activity that made history. Cold was nothing; man 'ne peut ni saisir ni communiquer' ('can neither grasp nor communicate') coldness.[49] Heat, however, conveyed both through experience over time, and through the force of Buffon's language, might be understood, translated and communicated through the outward movement of ideas as through that of molecules, allowing people to do what Nature no longer could.

Regulating heat meant being able to envision, in a gesture similar to Montesquieu's vision of political history discussed in Chapter 1, how all natural elements of an environment existed in relationship to one another. To alter the temperature of the Earth as a whole, people had to begin by understanding the mechanics of their own, local climate. Inhabitants of each type of climate would acquire knowledge of how nature behaved on this specific level, and then would put this particular knowledge into action in their immediate, familiar surroundings. To determine 'la température locale de chaque terre en particulier' ('the local temperature of each particular region'), one would consider first the amount of heat generated by every person, plant and animal – 'autant de petits foyers de chaleur' ('so many little heated hearths') – and then how these elements existed in 'proportion' to one another.[50] People and animals, for instance, 'répandent de la chaleur' ('emanate heat'), while plants (which do not move very much) 'ne produisent que de l'humidité froide' ('produce nothing but cold humidity'). If a region had more trees than people, for instance, the overall temperature of the climate would fall: 'Une seule forêt de plus ou de moins dans un pays suffit pour en changer la température: tant que les arbres sont sur pied, ils attirent le froid, ils diminuent par leur ombrage la chaleur du Soleil, ils produisent des vapeurs humides qui forment des nuages et retombent en pluie, d'autant plus froide qu'elle

48. Buffon, 'Des époques de la nature', p.244-46.
49. Buffon, 'Des époques de la nature', p.244.
50. 'Life thus burns […] with a sort of nonluminous flame. Animal fire differs only in degree from ordinary fire. It resembles ordinary fire in assimilating all matters that can serve it as aliment. The respiration of a little animal, for instance, absorbs as much as the light of a candle.' Hall, *Ideas of life and matter*, p.9-10.

descend de plus haut.'[51] Buffon used this same argument to explain why, even though the two cities had approximately the same latitude and elevation, Paris was so much warmer than Québec. He claimed there were fewer people in Québec to cut down all the forests there and thus to warm and cultivate the climate![52]

When people learned to judge the natural state of their climate, they could then change their relationship to what Buffon deemed to be the less productive elements of nature such as forests, and use them, against their innate qualities, to create heat: '[S]i ces forêts sont abandonnées à la seule Nature, ces même arbres tombés de vétusté pourrissent froidement sur la terre; tandis qu'entre les mains de l'homme, ils servent d'aliment à l'élément du feu, et deviennent les causes secondaires de toute chaleur particulière.'[53] Buffon's statement here may be extended to infer that humans, redirecting the forces of nature, were indeed for him the 'primary cause' and creators of the heat around them. When people had understood how nature operated to create a certain temperature, and when they had learned to actively modify this temperature, Buffon imagined that humankind entered into a state of happiness. Although it was conjectural and fictional, this was not an abstract concept. The trajectory of the 'Epoques de la nature' implied that happiness was the goal of a historical process: that of finding and recognizing what was sufficient and of learning to avoid extremes (be they of temperature, of violence, of ignorance). It was about deliberately making the temperate climate which allowed its inhabitants to overcome the vanity and egotism which historically caused humans to destroy one another for

51. 'One forest more or less in a country suffices to change its temperature: as long as trees are standing, they attract coldness, diminish the heat of the Sun with their shade, [and] produce humid vapors that form clouds and fall back to the earth in the form of rain which gets colder as it falls from a greater height.' Buffon, 'Des époques de la nature', p.243.
52. Buffon, 'Des époques de la nature', p.240. On the concept of degeneration, see Buffon's essay 'De la dégénération des animaux', in *Histoire naturelle* (1749-1767), vol.14, p.311-74, and Roger, *Les Sciences de la vie*, p.574-77. Buffon and Thomas Jefferson became embroiled in an argument on this subject, Jefferson contending that America and all its people lived in a thriving climate with large animals such as moose and mammoths. On this debate, see for example Lee Alan Dugatkin, *Mr. Jefferson and the giant moose: natural history in early America* (Chicago, IL, 2009).
53. '[I]f these forests are left to Nature, these same dilapidated, fallen trees rot coldly on the earth; while in the hands of man, they nourish the element of fire, and become the secondary causes of all local [forms of] heat.' Buffon, 'Des époques de la nature', p.243-44.

selfish reasons, and to reflect in an increasingly universal sense upon how to make their region, their country, their continent and finally the whole globe more peaceful, productive and beautiful. Happiness was a result of the temperate or the enlightened adjustment of extremes, just as the temperate was the result of the balancing of the 'expansive force' of impulsion with the 'attractive force' of gravity, just as science was the result of the equilibrium of the structure of human language with the structure of the world – an equilibrium that did not stagnate but rather opened up to new, not yet invented possibilities. When people understood how to induce this state of equilibrium, Buffon implied that they could not only delay the frozen end of the world and hinder the deterministic march of history, but also create temperance, a state in which future terrestrial paradises might be imagined and built. As Jean Svagelski has written of the Enlightenment vision of human progress and tolerance:

> Cet optimisme social se situe au croisement d'une chute de l'énergie naturelle et d'un progrès de la vie collective. Les encyclopédistes n'acceptent un tel progrès social qu'à la condition qu'il ne se marque pas par une simple urbanisation, mais qu'il représente un approfondissement de la raison et un accroissement du savoir [...]. L'homme ne troque plus son dynamisme primitif contre la sécurité d'une vie paisible, il transmute son énergie naturelle en énergie historique. Il ne s'assoupit pas dans le confort, il est tout entier tendu vers un progrès historique.
>
> This social optimism is situated at the crossroads of a decrease in natural energy and a progress of collective life. The encyclopedists only accepted such social progress on the condition that it was not marked by simple urbanization, but that it represented a deepening of reason and a growth of knowledge [...]. Man no longer trades his primitive dynamism for the security of a peaceful life; he transmutes his natural energy into historical energy. He does not waste away in comfort; he is completely open toward historical progress.[54]

Those living in a temperate climate of their own making embodied, for Buffon, one definition of the Enlightenment materialist and

54. The creative energy of history is produced both 'par la connaissance et par l'action' ('by knowledge and by action'). Svagelski, *L'Idée de compensation en France*, p.211-12. On Enlightenment history and its optimistic relationship to cultivation and climate change, see Wolloch, *History and nature in the Enlightenment*, p.83-92, 106-15, 116-22 and 243-54.

philosophe, whose mind was temperate and tolerant, open to modification, and whose written language in turn worked to modify the relationships between people and the spaces they inhabited. Buffon hoped his readers could learn to interact with the surrounding environment, with the goal of regulating temperature to produce peace and happiness: 'Heureuses les contrées où tous les élémens de la température se trouvent balancés, et assez avantageusement combinés pour n'opérer que de bons effets!' ('Happy are the countries where all elements of temperature are balanced and well enough combined to bring about nothing but good effects!').[55]

To conclude the 'Septième époque', Buffon reiterated that the more people acted, the more they learned and knew, and the more authorial power they had over the world:

> Tous ces exemples modernes et récens, prouvent que l'homme n'a connu que tard l'étendue de sa puissance, et que même il ne la connoît pas encore assez; elle dépend en entier de l'exercice de son intelligence; ainsi plus il observera, plus il cultivera la Nature, plus il aura de moyens pour se la soumettre et de facilités pour tirer de son sein des richesses nouvelles, sans diminuer les trésors de son inépuisable fécondité.
>
> All these recent and contemporary examples prove that man realized the extent of his power very late, and that even now he does not know it enough. This power depends entirely on the exercise of his intelligence; thus the more he observes, the more he cultivates Nature, and the more he will have the means to subdue it and the facilities to take new riches from its heart without diminishing the treasures of its inexhaustible fecundity.[56]

He suggested that history be read as a progression, or a process, toward happiness, and thus toward an ever more stable, temperate relationship between humans and nature:

> Ne nous arrêtons pas plus long-temps sur le triste spectacle de ces révolutions de mort et de dévastation, toutes produites par l'ignorance; espérons que l'équilibre quoiqu'imparfait qui se trouve actuellement entre les puissances des peuples civilisés se maintiendra et pourra même devenir plus stable à mesure que les hommes sentiront mieux leurs véritables intérêts, qu'ils reconnoîtront le prix de la paix et du

55. Buffon, 'Des époques de la nature', p.246.
56. Buffon, 'Des époques de la nature', p.253.

> bonheur tranquille, qu'ils en feront le seul objet de leur ambition, que les Princes dédaigneront la fausse gloire des conquérans et mépriseront la petite vanité de ceux qui pour jouer un rôle les excitent à de grands movemens.
>
> Let us not linger too long on the sad spectacle of these revolutions of death and devastation, all of them produced by ignorance; let us hope that the equilibrium, although imperfect, that is found today among the governments of civilized peoples will maintain itself and might even become more stable as people better understand their true interests, as they recognize the price of peace and tranquil happiness, and as princes disdain the fake glory of conquerors and scorn the petty vanity of those who, in order to take part, encourage these conquerors to commit dramatic acts.[57]

According to this citation, enlightened people, and especially enlightened monarchs, did not just reheat the world, but rather understood how to make the two extremes of hot and cold work together. Discovering this relationship and balance was an ever-adaptive process; it was based upon the constant activity of judging the relationship between the human perception of the world and the surrounding reality, and considering how to both perceive and act in this world better and more innovatively. By creating what Buffon called 'une Nature nouvelle', people could postpone the cold death of the world into the future, through the learned process of temperance – a process that was at once poetic, natural, historical and political.[58] The 'Epoques' might be seen as an example of what Nancy Struever calls 'a rhetoric of use'. To the importance in the eighteenth century of classical rhetoric – '*docere, delectare, movere*: rhetoric as a discipline regulated a discourse of instruction, delight, and persuasion' – was added a moral value, 'by locating it [rhetoric] in a pragmatic domain of use, of political or social utility'. There was a 'movement away from a rhetoric defined simply as "trivial" discipline and toward a rhetoric of use, and of use of argument, a rhetoric which claims civilized discussion as the proper domain for the exercise of civilized talents'.[59]

57. Buffon, 'Des époques de la nature', p.239.
58. This expression, 'a new Nature', is from Buffon, *Histoire naturelle* (1749-1767), vol.12, p.xii-xiii.
59. Nancy Struever, 'The conversable world: eighteenth-century transformations of the relation of rhetoric and truth', in *Rhetoric and the pursuit of truth: language change in the seventeenth and eighteenth centuries*, ed. Brian Vickers and Nancy Struever (Los Angeles, CA, 1985), p.79-119 (79, 83).

Indeed, in keeping with 'the exercise of civilized talents', the last sentence of the 'Epoques' reflected on how, after centuries of violence and ignorance, of wasted heat, 'enfin il a reconnu que sa vraie gloire est la science, et la paix son vrai bonheur' ('finally [man] has recognized that his true glory is science, and peace his true happiness'), understanding, through the dynamism of language, the dynamic relationship between his activity and that of nature.[60] Buffon wrote of this state of humans and nature as a hypothetical ideal. Yet it was an ideal that he seemed to think people were capable of imagining, and thus someday capable of acting for. It was as if Buffon hoped that the persuasive impetus of his text would push humans to invent and enact an 'Eighth epoch', the chapter in which they would think, act and write the story of a temperate Earth by changing the nature of the planet itself.

60. Buffon, 'Des époques de la nature', p.254.

Conclusion: preserving the heat of the *Histoire naturelle*

'Quel est le but de tout cela?'
'Peut-être qu'il n'y a pas de but.'
'Cependant…' et Pécuchet répéta deux ou trois fois 'cependant' sans trouver rien de plus à dire.
'N'importe, je voudrais bien savoir comment l'univers s'est fait.'
'Cela doit être dans Buffon', répondit Bouvard […].
Les *Epoques de la nature* leur apprirent qu'une comète, en heurtant le soleil, en avait détaché une portion, qui devint la terre. D'abord les pôles s'étaient refroidis. Toutes les eaux avaient enveloppé le globe; elles s'étaient retirées dans les cavernes; puis les continents se divisèrent, les animaux et l'homme parurent. La majesté de la création leur causa un ébahissement infini comme elle. Leur tête s'élargissait. Ils étaient fiers de réfléchir sur de si grands objets.

'What is the point of it all?'
'Perhaps there isn't a point.'
'Yet…' and Pécuchet repeated the word two or three times, without finding anything more to say.
'It doesn't matter, I should like to know how the universe came about.'
'It must be in Buffon,' answered Bouvard […].
The *Natural Epochs* told them that a comet striking the sun had knocked off a piece, which became the earth. First, the poles had cooled down. All the waters had covered the

> globe; they had withdrawn into hollows, then the continents divided off, animals and man appeared.
> The majesty of Creation caused them an awe as infinite as itself. Their heads expanded. They were proud of reflecting on such grand subjects.[1]

Rethinking Buffon's intellectual legacy

Scholars have struggled to make sense of the impact of Buffon's death and his legacy in the nineteenth century. His work resists placement in one modern disciplinary category or another. It is well known that his ideas about geology, Earth history and evolution influenced later work in these areas – although attention should be paid to ensuring that Buffon is not read anachronistically, as having foreshadowed or been a precursor to more advanced theories in the biological and geological sciences. It is important to reconsider the argument that, after his death, Buffon lost all scientific credibility because his ideas were considered outdated, and that therefore his influence was evacuated from the nineteenth-century scientific disciplines. The question of Buffon's influence, moreover, pertains not only to the natural sciences but also to the disciplines of history and literature which were themselves impacted by change in the larger structure and implications of scientific thought. Studying Buffon's effect on later generations of intellectuals needs to be focused on the question of how the languages of these disciplines expressed scientific ideas.

What was Buffon's influence on the natural sciences, in particular natural history, the community with which he identified during his life? A prevailing belief is that, upon Buffon's death, this area of study was finally able to shake off the weight of poetic, lyrical expression that hindered a denotative, precise understanding of the world. To argue that natural history during the French Revolution and the early nineteenth century sought to move away from the importance of language as a tool in the creation of natural knowledge is a generalization that misses how the ideas and prose of the *Histoire naturelle* were used by naturalists in the early 1790s. Pietro Corsi, for instance, contends that the generation of natural historians following Buffon

1. Gustave Flaubert, *Bouvard et Pécuchet* (Paris, 1945), p.80; *Bouvard and Pécuchet*, translated by A. J. Krailsheimer (New York, 1976), p.85-86.

'foresaw the end of natural history as a literary genre and a favorite arena for theoretical speculation and philosophical cosmogony. In its place, they would introduce rigorous procedures for naturalistic investigation, modeled on the logical rigor of Linnaean taxonomy and the epistemological criteria of Lockean origin, reworked in France by Condillac, his disciples, and his followers.'[2] Corsi sees further proof for the divergence from older forms of natural history in the creation of more specialized disciplines, each with its specific, precise vocabulary.[3] Yet Condillac's epistemology and Linnaeus' taxonomy could not have been more deeply founded in the Enlightenment project of translating reason into written languages that, their authors believed, could mirror the order of nature. Theirs certainly was a more denotative order than Buffon's, but it was still intrinsically tied to language. It was not about history, becoming and hypothetical invention, but rather about discovering a perfect transparency between words and natural things that could ultimately efface complexity and bring about a perception of the world as a series of general, simple *rapports*. Condillac's philosophy was used during the early Revolution to reform both science and politics. One only need look at Antoine Lavoisier's essays and tables on chemical nomenclature, in which it was equally important to identify the nature and purpose of elements entering into chemical reactions as it was to find the true, natural names for each of these elements. Far from becoming separated, language and the phenomena it signified remained interchangeable.[4] In other words, even though Buffon's model for natural history was not the prevailing one during the Revolution, but his understanding of science as the linking of words and natural things was shared by many thinkers. The search for a language of nature did not simply disappear, even if the terms and methods changed.

The language of the *Histoire naturelle* itself was not, moreover, immediately rejected. As Emma Spary has shown, Buffon's language was used in the early years of the Revolution as a source of identity and unity among the naturalists at the Jardin and Cabinet du roi,

2. Corsi, *The Age of Lamarck*, p.4.
3. Pietro Corsi, 'Buffon sous la Révolution et l'Empire', in *Buffon 88*, ed. J.-C. Beaune *et al.*, p.639-48 (642).
4. See the *Méthode de nomenclature chimique* (Paris, Cuchet, 1787), by Morveau, Lavoisier, Bertholet and de Fourcroy, which included the *Tableau de la nomenclature chimique*. On Lavoisier, nomenclature and Revolutionary social and political reforms, see Wilda C. Anderson, 'Scientific nomenclature and Revolutionary rhetoric', *Rhetorica: a journal of the history of rhetoric* 7.1 (winter 1989), p.45-53.

which would become the Muséum d'histoire naturelle. In a revolutionary society where older systems of patronage and intellectual institutions were dissolving or their political roles being questioned, the importance and function of natural history with respect to the sciences, the state and education became a matter of debate. Spary describes how many of the natural historians at the Jardin du roi gathered around the memory of Buffon in order to reassert, strengthen and redefine a community of thought.[5] Seeking funding for the survival of the Jardin, its members presented a speech to the Assemblée nationale in 1790: the 'Première adresse des officiers du Jardin des plantes et du Cabinet d'histoire naturelle'. This address borrowed from Buffon's ideas about the noble, ethical relationship between nature and the natural historian, and even tried to imitate his writing style.[6] To quote from the speech, Daubenton and his peers there wrote:

> [O]n ne s'est pas contenté d'y réunir des divers points du globe les dépouilles préparées avec soin de presque tous les animaux connus, et d'y offrir à l'étude une suite complète de toutes les sortes de minéraux que l'on a découverts jusqu'à présent: mais tous ces objets y sont indiqués par des inscriptions étendues, de telle sorte que les galeries qui les renferment sont comme un grand livre ouvert, où [...] la nature parle, pour ainsi dire, et s'exprime elle-même; elle y montre, en quelque sorte, les productions qui se ressemblent et celles qui diffèrent; [...] les minéraux qui peuvent servir de signes pour en faire reconnoître d'autres; [...] le pouvoir de l'homme pour perfectionner ou maintenir les races et acclimater les espèces éloignées; l'histoire enfin des tems et des lieux, écrite, pour ainsi dire, par sa main toute-puissante.
>
> We were not content to reunite [in the museum] the remains, carefully prepared, of all known animals of the different parts of the globe, and to offer for study a complete chain of all the types of minerals we have discovered to the present day; all of these objects are [furthermore] identified by long inscriptions such that the galleries that house them are like a great open book where [...] we might say that nature speaks and expresses herself; she shows here, in a way, the productions that resemble one another and those that differ; the minerals which can serve as signs to identify others; [...] the power of mankind to perfect

5. Emma C. Spary, *Utopia's garden: French natural history from Old Regime to Revolution* (Chicago, IL, 2000), p.157-58 and 173-90.
6. Spary, *Utopia's garden*, p.166-67.

> or maintain races and acclimate distant species; the history, finally, of time and spaces written, so to speak, by her [nature's] all-powerful hand.[7]

The language here emulates Buffon's way of writing and his ideas. The address describes the museum and gardens at the Jardin as a microcosm of the larger world, brought into existence and motion by the interrelations between natural activity and human thought and writing. Not only were natural specimens reunited in one space, but they were accompanied by 'inscriptions' which linked them to one another and to the more general language of *rapports* of which the book of nature was composed.

Using Buffonian language to promote the value of natural history to a new political regime contributed to the naturalist's immortalization as one of the great intellectual figures in France: the father of natural history, the ancestor figure from which the next generation of naturalists drew its legitimacy. The Muséum was one of the only royal institutions to survive the Revolution. Spary claims that its survival was largely due to the fact that its faculty adopted Buffon's rhetoric, especially clear in the 'Epoques de la nature', of the interrelated scientific and moral trajectories of humans and nature.[8]

Buffon's work also shaped the discipline of literature. In fact, his presence in this domain looms so large that, paradoxically, it has been taken for granted and often forgotten. In his 'Avant-propos' to the *Comédie humaine*, Honoré de Balzac revealed the influence of Buffonian philosophy when he argued that: 'L'histoire n'a pas pour loi, comme le roman, de tendre vers le beau idéal. L'histoire est ou devrait être ce qu'elle fut; tandis que *le roman doit être le monde meilleur*, a dit Mme Necker [...]. Mais le roman ne serait rien si, dans cet auguste mensonge, il n'était pas vrai dans les détails.'[9] This quotation

7. Louis-Jean-Marie Daubenton *et al.*, 'Première adresse des officiers du Jardin des plantes et du Cabinet d'histoire naturelle', reprinted in Ernst-Théodore Hamy, 'Les derniers jours du Jardin du roi et la fondation du Muséum d'histoire naturelle', in *Centenaire de la fondation du Muséum d'histoire naturelle, 10 juin 1793-10 juin 1893: volume commémoratif publié par les professeurs du Muséum* (Paris, 1893), p.97-100 (98). This work came to my attention in Spary, *Utopia's garden*, p.166-67.
8. Spary, *Utopia's garden*, p.190-91.
9. 'It is not a law to history, as it is to romance, to make for a beautiful ideal. History is, or ought to be, what it was; while romance ought to be "the better world," as was said by Mme. Necker [...]. Still, with this noble falsity, romance would be nothing if it were not true in detail.' Honoré de Balzac, 'Avant-propos', in *La Comédie humaine*, vol.1 (Paris, 1842), p.7-32 (23; original emphasis).

(partially attributed to Mme Necker, one of Buffon's great allies) hints at Buffon's literary legacy in nineteenth-century France by following the idea of using history to imagine a better world. In Balzac's project to combine natural law with the laws of societies and human behavior, he cited the names and ideas of several important thinkers of the long Enlightenment: not only Buffon but also Swedenborg, Leibniz, Goethe, Charles Bonnet, Needham, Cuvier and Geoffroy Saint-Hilaire. His interest in the famous 1830 debate on animal structure and origin between the last two figures on this list, Cuvier and Saint-Hilaire, derived from his understanding of the scientific terms of their quarrel from within a hypothetical structure. It was not about what they were saying, but how they ordered and articulated their ideas into systems, potential worlds generated from thought. He wrote:

> Ce serait une erreur de croire que la grande querelle qui, dans ces derniers temps, s'est émue entre Cuvier et Geoffroi Saint-Hilaire, reposait sur une innovation scientifique. L'unité de composition occupait déjà sous d'autres termes les plus grands esprits des deux siècles précédents. En relisant les œuvres si extraordinaires des écrivains mystiques qui se sont occupés des sciences dans leurs relations avec l'infini, tels que Swedenborg, Saint-Martin, etc., et les écrits des plus beaux génies en histoire naturelle, tels que Leibnitz, Buffon, Charles Bonnet, etc., on trouve dans les monades de Leibnitz, dans les molécules organiques de Buffon, dans la force végétatrice de Needham, dans l'emboîtement des parties similaires de Charles Bonnet, assez hardi pour écrire en 1760: L'animal végète comme la plante; on trouve, dis-je, les rudiments de la belle loi du soi pour soi sur laquelle repose l'unité de composition.

> It is a mistake to suppose that the great dispute which has lately made a stir, between Cuvier and Geoffroy Saint-Hilaire, arose from a scientific innovation. Unity of structure, under other names, had occupied the greatest minds during the two previous centuries. As we read the extraordinary writings of the mystics who studied the sciences in their relation to infinity, such as Swedenborg, Saint-Martin, and others, and the works of the greatest authors on Natural History – Leibnitz, Buffon, Charles Bonnet, etc., we detect in the monads of Leibnitz, in the organic molecules of Buffon, in the

---

The English translation is from 'Author's introduction', translated by George Saintsbury, in *The Works of Honoré de Balzac*, ed. William P. Trent (New York, 1900), p.li-lxvii (lxi).

> vegetative force of Needham, in the correlation of similar organs of Charles Bonnet – who in 1760 was so bold as to write, 'Animals vegetate as plants do' – we detect, I say, the rudiments of the great law of Self for Self, which lies at the root of Unity of Plan.[10]

According to this quote, the implications of the Cuvier–Saint-Hilaire debate were not based on an 'innovation scientifique' but rather on the search for an all-encompassing, harmonizing worldview prominently visible in the writings of Enlightenment philosophers. Through language they sought the law that could explain the *rapports* unifying the world: 'l'unité de composition' which would establish a seamless flow between minerals, plants, animals, humans and the heavens. This was not simply a way to explain the operation of the world; it was a model for the 'monde meilleur' of the first quotation, a better world made possible through the organizational and creative power of human thought over the facts and details of scientific reality.[11] The result, for Balzac, was a novel: 'a noble falsity, [...] true in detail'.

Balzac understood the novel as the discursive space in which a unifying worldview could be envisioned and its ability to interpret and structure reality proven. Buffon was one of the authors who enabled Balzac to attribute such powers to literature and in particular to the novel. Although written language was the force behind Buffon's knowledge and ordering of nature, and although he used aspects of literary, lettered culture to build the discipline of natural history, he would not have referred to his work as literature. Yet Buffon was an important figure for this developing and changing area of study. He was a substantial – if often silent and unquoted – part of the fabric of the discursive reality of authors such as Balzac who explicitly questioned older meanings of literature and carved out a space of literary invention that relied upon the tools once used to understand

10. Balzac, 'Avant-propos', p.8-9; 'Author's introduction', p.li-lii.
11. Peter Hanns Reill has demonstrated Buffon's influence on German 'Natur-Philosophie' in the early nineteenth century. Scholars were inspired by Buffon's seamless chain of living matter that stretched across time, telling the history of nature's past and redirecting its future. See Reill, 'History and the life sciences at the beginning of the nineteenth century: Wilhelm von Humboldt and Leopold von Ranke', in *Leopold von Ranke and the shaping of the historical discipline*, ed. Georg G. Iggers and James M. Powell (Syracuse, NY, 1990), p.21-35 (22-23); and 'Buffon and historical thought in Germany and Great Britain', in *Buffon 88*, ed. J.-C. Beaune *et al.*, p.667-79 (672-73.)

the natural world. What Buffon had seen as part of natural history – *style, belles-lettres, critique* – they incorporated into literature.

It became possible to see Buffon's work as literary after the publication of the 'Epoques de la nature', reviewed in the *Correspondance littéraire* as 'un des plus sublimes romans, un des plus beaux poèmes, que la philosophie ait jamais osé imaginer' ('one of the most sublime novels, one of the most beautiful poems that philosophy ever dared imagine').[12] This was not to be understood as praise. Buffon would not have used the term 'roman' to describe his work. He would have considered the capacity of his language and style to build parallel worlds more beautiful and poetic than nature itself, as part of the scientific work of both simulating and recreating the energy and heat of the world. However, many of the critiques of the 'Epoques' did not appear to have understood Buffon's work as a scientific activity. He was increasingly critiqued on the basis that many of the facts he used, in particular those that helped to support the story of the creation of the planets and those that explained the development and movements of species over time, were now deemed incorrect or were thought to have been discovered through outdated methods (although it must not be forgotten that Buffon was one of the first scholars to publish an age of the Earth far older and more radical than those commonly accepted; and the first to establish an enduring law about the distribution of animal species, the so-called 'Buffon's law').[13] To state that Buffon wrote novels and poetry was to reveal how far removed his work had come to be considered from the study of the real, phenomenal world.

Although Buffon was criticized at the end of his career for writing 'sublime novels' when he believed he was doing quite the opposite, it is precisely through the modern study of literary analysis that the *Histoire naturelle* as a form of natural science ultimately may be understood. These tools show how language and knowledge of nature were synonymous throughout much of the Enlightenment. By interpreting Buffon's work as a literary endeavor, as I have done in this book, I hope to have more accurately shown its contemporary value as a serious scientific object. Knowing was about finding the language that best expressed the processes of both human thought and nature. Studying the natural world and creating a science of natural history was an exercise in perception, representation and communication.

12. Quoted in Roger, *Buffon*, p.554.
13. Roger, *Buffon*, p.555. On the reception of the 'Epoques de la nature', see Roger, 'Introduction', p.cxxix-cxliv.

Reading the *Histoire naturelle* through available contemporary interpretations of both letters and natural knowledge does not detract from the seriousness of the project, and in fact produces a more complete conception of the ideas and endeavors of Enlightenment rationality.

The preceding chapters have provided examples of this rationality at work. They have examined how Buffon used stylistic, discursive devices, many of which today would be called literary, but which in the Enlightenment were tools for the articulation of human reason and the intellectual processes through which human ideas were brought into proportion with the order of nature. The organization and transmission of thought and reason through style and heat made it possible to write history and to hypothesize or invent new possibilities. From the foundations of rationality, Buffon believed he could change not only the nature of a reader's perception but indeed the nature of the world itself. Buffon's work may be understood as a form of science because it imparted a literacy of nature, which allowed the Earth to be read and ordered as a language and then remade, remodeled and controlled through this language. Knowing the world through language and inventing a new nature from it was considered a valid and effective way of practicing natural history.

Buffon's work embodies the Enlightenment paradigm that language and the reimagination and reconception of reality in a parallel, possible textual world could impart to readers the seeds of action and intervention in the real world. This was the belief that Balzac adapted to the form of the novel. As disciplinary boundaries and definitions began to shift, the 'sublime novels' and 'beautiful poems' that spoke of nature became more distinctly part of literature and history. And yet, the creative, inventive capacities of written language, the discursive models discussed in this book, continued to have important epistemological objectives. Buffon understood natural science as a language with different modes for creating different types of relationships between humans and the world. The ideas of style and hypothesis allowed world-making – be it fictitious or reconstructions of the past – to be used not as an escape from reality, but rather as a way to rethink and modify reality's very structure.

Condorcet's *Eloge de M. de Buffon*

This study opened with an analysis of how Buffon worked within existing categories of knowledge in the 1740s to modify the discipline of natural history. It closes with a return to investigating and expanding upon the relationship between literature and knowledge by analyzing another moment of disciplinary foment and change. Understanding the impact of the discipline of natural history as reformed by Buffon, and looking forward to how Buffon's language of nature influenced literature, necessitates consulting a text that embodies the process of this paradigm shift. The marquis de Condorcet's *Eloge historique de M. le comte de Buffon*, read before the Académie royale des sciences, provides an inside look at how Buffon's ideas, while remaining quite influential, became less predominant in the natural sciences and more useful to literature. This text captures the ways in which the study of the natural world at the end of the eighteenth century moved away from Buffon's understanding of it. It reveals the changes and transitions that occurred in the perception of the naturalist's work by a subsequent generation of intellectuals, and shows how these transitions were articulated in discourse. Interestingly, despite a growing divide between knowledge of the natural world and the uses and applications of probable, parallel, literary worlds, these questions may be approached through the same principles of Buffon's language analyzed throughout this book, notably style and heat.

After Buffon's death in April 1788, Condorcet, in his role as secretary of the Académie royale des sciences, composed Buffon's eulogy, pronouncing it to his peers on December 12, 1788. Condorcet's interpretation of Buffon's work is an example of a profound reflection upon the meaning and function of late-Enlightenment disciplinary boundaries between natural knowledge and literature. What a new generation of thinkers considered scientific at the end of the eighteenth century was not the same as what Buffon had believed, and Condorcet attacked the science and methods of the *Histoire naturelle* with biting sarcasm and irony. Yet, despite his critique, or perhaps because of it, in the *Eloge de M. de Buffon* he seemed to sense that Buffon's oeuvre had to be rescued from potential obscurity. Ultimately the irony in Condorcet's writing was not so much destructive as constructive, a figure of style which allowed him to carve from within an older paradigm a new way in which to understand the legacy of Buffon's achievements.

The way in which Condorcet read Buffon's ideas in his *Eloge* would have been unrecognizable even a decade earlier. And yet he established criteria according to which the *Histoire naturelle* could remain readable henceforth, and these criteria were distinctly literary. He often made use of the same terminology as Buffon's generation of intellectuals; however, it no longer carried the same meanings and functions.

Condorcet did not think like Buffon, but his thought imparted a more recognizable vision of modern disciplinary distinctions between science and literature. In the *Eloge*, Buffon's voice lived on through his style – both through his individual, personal voice and through the qualities of the Enlightenment philosophies that foregrounded his thought. Heat was returned to its status as a metaphor and moreover was dissociated from scientific activity. And yet language and its ability to reproduce the vitality of nature remained essential, teaching how to modify perception and to change how reality was seen and lived, activities now often identified with the purpose and power of literature.

Condorcet's *Eloge* of Buffon may be understood as an epistemological exercise. In it, he asked his public to reflect upon what he perceived to be the changing status of approaches to natural philosophy and the resulting disciplinary boundaries. The critical tone of his eulogy can be read as an instance of the development of an understanding of scientific activity very different from that of Buffon. It seems that Condorcet – and probably a growing number of his colleagues – no longer understood Buffon's ideas as scientific, and did not appreciate the florid style in which they were written and to which they owed their very essence. He described Buffon's exaggerated and inflated rhetoric, mirrored in the infamous egotistical character of the naturalist.[14] Mockingly imitating the expressive, energetic use of heat in Buffon's 'Epoques de la nature', Condorcet wrote that the natural historian 'ne vit que la nature, à la fois riante, majestueuse & terrible, offrant des asyles voluptueux & de paisibles retraites entre des torrens de laves & sur les débris des volcans, prodiguant ses richesses à des campagnes qu'elle menace d'engloutir sous des monceaux de cendres ou de fleuves enflamés'.[15] Condorcet raised the question of whether this

14. The *Eloge de M. de Buffon* was not the only venue in which Condorcet criticized Buffon's style, writing and character. In his correspondence with Turgot and Voltaire, Condorcet did not speak favorably of Buffon, complaining about the annoyances the naturalist caused in his professional life, as well as making fun of Buffon's vanity and outlandish theories of the history of the Earth. See Elisabeth Badinter and Robert Badinter, *Condorcet (1743-1794): un intellectuel en politique* (Paris, 1988), p.119-20.

15. '[S]aw only nature, at once pleasant, majestic and terrible, presenting voluptuous refuges and peaceful sanctuaries among torrents of lava and upon the debris of volcanoes, dispensing its riches upon lands that it threatens to engulf under mounds of ash or burning rivers.' Marie-Jean Antoine Nicolas, marquis de Condorcet, *Eloge historique de M. le comte de Buffon* (Paris, Aux Deux Ponts, 1792), p.6.

extravagant description of a nature both harsh and inviting to people in fact revealed anything about how the world actually functioned. It is especially interesting that he ridiculed Buffon's interest in heat – volcanoes, lava, ashes, boiling water – and the creative dominance it gave to nature. In so doing, Condorcet contested both the idea that the history of nature was one of dramatic, impetuous forces and that this world could be portrayed and known through a singular, commanding authorial power. Condorcet distinguished the energetic language of heat from the production of knowledge; instead of fashioning a parallel world considered as true, if not more true, than nature itself because of the liveliness of its discourse, representation now indicated moving farther away from reality.[16]

Condorcet, usually a prolific writer, supposedly had to write and rewrite the section of the *Eloge* about Buffon's style seven times.[17] This detail alone is a clue that, despite his ridicule of Buffon's language, and despite the fact that he did not see it as scientific, the importance and influence of language and style retained its grasp and remained a significant issue for Condorcet. Buffon's writing embodied the mind of the author himself, transmitting an enduring discursive energy and making it difficult to escape his famed words: 'le style est l'homme même' ('style is man himself').[18] The *Eloge de M. de Buffon* was thus necessarily about language and its epistemological – because discursive – power. It was as much about reflecting upon older ways of thinking and conveying knowledge, as it was about reconsidering the relationship between observers and nature to create and present a revised understanding and language of scientific activity. That the activity of composing such a eulogy was deeply tied to style, as Buffon

16. Condorcet wrote: 'La difficulté de répandre de l'intérêt sur tant d'objets inanimés ou insipides ne l'arrêta point; il avoit déjà cette conscience du talent qui, comme la conscience morale, ne trompe jamais quand on l'interroge de bonne foi, & qu'on la laisse dicter seule la réponse.' ('The difficulty of spreading interest over so many inanimate or insipid objects never stopped him; he already had that awareness of talent which, like moral conscience, never deceives when examined in good faith and when one allows it to alone dictate the reply.') Condorcet, *Eloge historique*, p.13-14.
17. Elisabeth and Robert Badinter note here that 'rarement Condorcet s'était donné autant de mal pour faire un éloge, lui qui écrivait d'un trait et corrigeait peu son premier jet' ('rarely did Condorcet struggle so much to write a eulogy, he who wrote in a single stroke and made few corrections to his first draft'); Badinter and Badinter, *Condorcet*, p.243.
18. Buffon, 'Discours sur le style', p.11.

would have understood it, was ironically and unavoidably essential to Condorcet's presentation.

In his *Eloge historique* Condorcet introduced Buffon's *Histoire naturelle* with these words:

> En parlant de cet ouvrage que tous les hommes ont lu, que presque tous ont admiré, qui a rempli, soit par le travail de la composition, soit par des études préliminaires, la vie entière de M. de Buffon, nous ne prendrons pour guide que la vérité (car pourquoi chercherions-nous vainement à flatter par des éloges qui ne dureroient qu'un jour, un nom qui doit vivre à jamais?) & en évitant, s'il est possible, l'influence de toutes les causes qui peuvent agir sur l'opinion souvent passagère des contemporains, nous tâcherons de prévoir l'opinion durable de la postérité.
>
> When speaking of this work that everyone read, that almost everyone admired, that occupied, both through the work of composition and through preliminary studies, the entire life of M. de Buffon, we will take only truth as a guide (for why would we vainly seek to flatter in eulogies, which only last a day, a name which must live on forever?) and by avoiding, if possible, the influence of anything that could influence the fleeting opinions of contemporaries, we will try to foresee the lasting opinion of posterity.[19]

Certainly this passage was sarcastic. Flattering Buffon would be to no avail, for Condorcet had nothing flattering to say. In any case, he argued, flattery would only reflect the opinions, whims and prejudices of the moment and not the enduring truth of posterity. Condorcet sought the 'truth', using the eulogy as a means of determining what aspects of Buffon's writing and thought would stand the test of time, and why they would last. Whether or not he and his contemporaries liked it, Buffon's was a name 'which must live on forever'. Condorcet had therefore to find the mechanisms, both epistemological and discursive, to define the nature of the *Histoire naturelle* for posterity and to define its future life.

Condorcet began his speech by mocking the idea of truth in the *Histoire naturelle*, that is, the assumptions and processes through which readers were meant to accept Buffon's facts as scientific. As he disproved the facts of the *Histoire naturelle* and showed how they did not correspond to reality, and as he denied Buffon's practice of re-presenting nature through painting with words as a positive

19. Condorcet, *Eloge historique*, p.14.

activity of knowledge-making, Condorcet simultaneously revealed the continuing intellectual impact of Buffon's work. It was not, he wrote, the conviction that readers received from representations of nature in the *Histoire naturelle* that made this work informative and useful. Rather, he decided, it was Buffon's errors, both implicit and unconscious, that pushed readers in a quest to perfect their judgment and to seek true knowledge of nature:

> Mais M. de Buffon connoissoit trop les hommes pour ne pas sentir qu'une science qui n'offriroit que des faits particuliers, ou ne présenteroit des résultats généraux que sous la forme de simples conjectures, frapperoit peu les esprits vulgaires, trop foibles pour supporter le poids du doute. Il savoit que Descartes n'avoit attiré les hommes à la philosophie que par la hardiesse de ses systèmes, qu'il ne les avoit arrachés au joug de l'autorité, à leur indifférence pour la vérité, qu'en s'emparant de leur imagination, en ménageant leur paresse, & qu'ensuite libres de leurs fers, livrés à l'avidité de connoître, eux-mêmes avoient su choisir la véritable route.

> But M. de Buffon understood people too well to not feel that a science that could only offer specific facts, or that could only present general results in the form of simple conjectures, would not really touch vulgar minds, too weak to bear the weight of doubt. He knew that Descartes had only attracted people to philosophy through the boldness of his systems, that he had only torn them from the yoke of authority and from their indifference to truth by taking possession of their imagination, by restraining their laziness, and thus freed from their chains, given over to the greed of knowing, they were themselves able to choose the true path.[20]

All the while mocking Buffon for appealing to a more general, less specialized public, and for attracting these readers through 'imagination' and not reason, Condorcet recognized the fact that the natural historian deliberately used persuasive language and sometimes even outdated facts to combat older ways of thinking and belief systems that could not handle the 'weight of doubt'. In a manner similar to Condorcet's own conception of error in the *Esquisse d'un tableau historique des progrès de l'esprit humain*, Buffon himself conceived of error as a positive, heuristic tool. He strategically used facts he did not hold to be true in his work, especially when presenting the controversial age of the Earth in the 'Epoques de la nature'. In the 'Première époque',

20. Condorcet, *Eloge historique*, p.15-16.

Buffon explained that, although he believed the Earth to be older than 75,000 years (an already radical figure for the time), he published this number so that readers would not be too shocked and reject his research outright. He held that this figure, although overstated, was not outside the limits of verisimilitude. It might initially anger or confuse readers, but would have enough persuasive strength to draw them in and make them read further. Buffon's proofs (published at the end of the 'Epoques') would slowly show readers how this figure could be understood as possible, then probable, and perhaps one day certain and accepted as true. Once his public came to terms with this number, he could then present a larger figure, and the persuasive process would begin anew. In this way, Buffon built into his text a mechanism for bringing his readers' knowledge into proportion with reality.[21]

Returning to Condorcet's critique, however, once able to accept Buffon's 'facts' and their place within his story of nature, readers could begin to see the faults of this system and its lack of correspondence with reality. Thus Buffon's conjectures gave readers a base from which to question and discover better 'preuves' and to subject 'les faits reçus' ('received facts') to 'un examen rigoureux' ('a rigorous examination'): 'ayant épuisé ces premières armes, on cherche de nouveaux faits pour s'en procurer de plus sûres & d'une trempe plus fine' ('having exhausted these first weapons, one looks for new facts in order to procure ones that are more certain and of a more refined constitution').[22] Once Buffon's readers began this process of improving and honing their understanding of nature, Condorcet sarcastically implied, they would soon recognize the faults in Buffon's own conception of the world. They would use Buffon's idea of error against him, and, through the activity of correcting his errors, they would implicitly continue their study of natural history.

The danger in Buffon's work, according to Condorcet, was the deliberate conflation of the rhetorical and epistemological functions of error. In his critique of Buffon, Condorcet separated the acquisition of knowledge and the ability to judge its truth value from the magnetic pull of persuasive language. Persuasive language could be deceptive, leading readers to believe that they held the truth without having studied what Condorcet considered to be the empirical and mathematical proofs requisite for reaching scientific knowledge of the

21. For more on Buffon's use of error in the 'Epoques', see Roger, 'Introduction', p.lxvi-lxvii, and Anderson, 'Error in Buffon', p.691-701.

22. Condorcet, *Eloge historique*, p.16.

world. He wrote that the 'admiration' for Buffon as a great author 'dispose si facilement à la croyance, que les lecteurs entraînés à la fois par le nom de l'auteur & par le charme du style, cédent sans résistance, & semblent craindre que le doute, en affoiblissant un enthousiasme qui leur est cher, ne diminue leur plaisir' ('Admiration so easily induces belief, that readers led at the same time by the name of the author and by the charm of style yield without resistance and seem to fear that doubt, by weakening the enthusiasm so dear to them, [would] diminish their pleasure').[23] For Condorcet, the real 'pleasure' derived from reading a scientific text was an active fortification against 'the charm of style', or the confusion of discourse with knowledge. Pleasure was not derived from emotion or imagination, from feeling and knowing nature through the energy and heat of prose. It was based in reason (still itself an interaction between thought and the world), in understanding how the mind generated relationships between ideas and in tracing how thought moved, step by step, toward more abstract and general conclusions. Its effects could be measured in the production of new forms of truth. Furthermore, pleasure for Condorcet embodied the act of providing human reason with a limit from which to measure its movement, to judge the certainty and utility of knowledge in any subject or domain, and to decide upon the truths of sciences based upon what was currently known and how it was known.[24] In a sarcastic sense, he derived pleasure from reading the *Histoire naturelle* because this text established a negative limit, away from which scholars could measure how far their scientific thought had come since the epoch of Buffon.

However, in another sense, the use of error to make readers constantly refine or replace the worlds they perceived or invented could still produce pleasure regardless of whether these worlds approached mathematical truth. Buffon's writings might not have been scientific

23. Condorcet, *Eloge historique*, p.29-30.

24. This limit is created through the mathematical calculation of probabilities, which Condorcet thought came closest to complete certainty. Humans could not perceive absolute truth. They must, as Keith Baker explains, 'approximate that knowledge [...] by subjecting the probable truths of experience to the precise evaluation of the calculus of probabilities'. Quoting from one of Condorcet's manuscripts on probability, Baker writes that this 'calculus of probabilities' provided for 'the moral and physical sciences together "the appropriate degree of precision and exactness which men may flatter themselves to attain."' See Keith Baker, *Condorcet: from natural philosophy to social mathematics* (Chicago, IL, 1975), p.180.

in Condorcet's opinion, but he must not have completely rejected their ability to elucidate or enlighten – especially as his passionate project, the 1795 *Esquisse d'un tableau historique des progrès de l'esprit humain*, was based upon an imagined narrative of the development of human civilization and ideas. Condorcet explained that '[l]orsque des opinions qui paroissent erronées se trouvent dans un livre fait pour séduire l'esprit comme pour l'éclaircir, c'est presque un devoir d'avertir de les soumettre à un examen rigoureux' ('When opinions that seem erroneous are found in a book written to seduce the mind as to elucidate it, it is almost a duty to subject them to a rigorous examination').[25] As such, even if Buffon's depictions of exploding volcanoes and boiling rivers of lava were in fact imaginary and even fantastical, the force and energy of the vibrant language in which they were depicted nevertheless was capable of repainting a story of nature in the reader's mind. It was not pointless, frivolous prose; from there, the reader could assess and modify this representation, refining it to become more rigorous and coherent with respect to contemporary knowledge. Although not for the same reasons as the scientific study of the real world, invented worlds thus also required the human mind to engage and to act. According to Condorcet, Buffon contributed

> au progrès d'une science qui, soumettant au calcul les événemens dirigés par des lois que nous nommons irrégulières, parce qu'elles nous sont inconnues, semble étendre l'empire de l'esprit humain au-delà de ses bornes naturelles, & lui offrir un instrument à l'aide duquel ses regards peuvent s'étendre sur des espaces immenses, que peut-être il ne lui sera jamais permis de parcourir.

> to the progress of a science that, by subjecting events driven by laws that we name irregular, because they are unknown to us, to calculation, seems to extend the empire of the human mind beyond its natural limits, and offers it an instrument through which its gazes might stretch out over immense spaces that perhaps it might never have been able to cross.[26]

This science, which Buffon called natural history, became the science of literature and its 'instrument' was written language. Deployed through error, style and heat, language remained the bridge between Buffon's worlds and reality.

25. Condorcet, *Eloge historique*, p.29.
26. Condorcet, *Eloge historique*, p.30.

Saving style for posterity

Condorcet shared with Buffon the belief that style was the element of a work which would last and remain useful to future generations. Parts of Condorcet's *Eloge* even resonated with the language of Buffon's 'Discours sur le style'. In some parts, Condorcet even made reference to the 'Discours sur le style'. Passages from each text are compared here:

Buffon, 'Discours sur le style':	Condorcet, *Eloge historique de M. le comte de Buffon*:
[L]es connoissances, les faits et les découvertes s'enlèvent aisément, se transportent, et gagnent même à être mises en œuvre par des mains plus habiles [...]. [C]ar il n'y a que la vérité qui soit durable et même éternelle. Or un beau style n'est tel en effet que par le nombre infini des vérités qu'il présente.	[L]es déclamations, les systèmes passent, & les faits restent; ces livres qu'on a surchargés d'ornemens pour les faire lire, seront oubliés; mais s'ils renferment quelques vérités, elles survivront à leur chute.
[K]nowledge, facts and discoveries are easily removed and altered, and even profit from the touch of more skillful hands [...]. [F]or there is only truth that is lasting and even eternal. Now, a beautiful style is nothing but the infinite number of truths that it presents.[27]	[D]eclamations and systems pass, and facts remain; those books that were overloaded with ornamentation to make them be read will be forgotten; but if they enclose some truths, these will survive their [the books'] downfall.[28]

These two authors did not have the same understanding of 'truth', 'error', 'pleasure' or what information would stand the test of time. Yet it seems that they would have agreed that the truth that would prevail for posterity was the presence or absence of style in a text. Condorcet, in keeping with the sarcastic tone of the *Eloge*, referred in the above passage to the work of style somewhat negatively and with less vehemence than Buffon, and yet, by recuperating the parts of the natural historian's work that he thought would (and did) last, he was

27. Buffon, 'Discours sur le style', p.10-11.
28. Condorcet, *Eloge historique*, p.36.

indeed doing as Buffon believed he himself had done through style. In the *Eloge* Condorcet wrote that style came from those authors 'dont les ouvrages produisent sur les hommes dispersés des effets plus répétés & plus durables' ('whose works produce in people everywhere the most repetitive and durable effects'), and 'dont les ouvrages ont changé la pente des esprits, les ont portés à une étude utile, & ont produit une révolution qui peut faire époque dans l'histoire des sciences' ('whose works changed the inclination of minds, led them to useful studies, and created a revolution that can be an epoch in the history of sciences').[29] Such works would become monuments in the history of knowledge, and the reading experience of these monuments would serve as a heuristic tool for '[le] perfectionnement de l'espèce humaine' ('[the] improvement of the human species') – which was what Buffon had hoped to accomplish in the 'Epoques de la nature' through the conquering of nature. Furthermore, Condorcet thought that these authors, '[ces] hommes éloquens nés avec le talent de répandre la vérité ou d'exciter le génie des découvertes [...] mériteront d'être placés au niveau des inventeurs' ('eloquent men born with the talent to spread truth or to inspire the genius of discoveries [...] deserve to be placed on the level of inventors').[30] As 'inventors', they followed the rules of stylistic invention to build persuasive and poetic arguments, using language to invent new, and often better, conceptions of reality.

To answer the question of how Buffon's style would persevere, Condorcet first described the writers he did not believe would be remembered for their style. In the case of authors like 'Despréaux, Racine, Fénelon, Massillon, Voltaire', he contended that '[o]n peut sans danger les prendre pour modèles' ('[o]ne can safely take them for models'), because 'le grand secret de leur art est de bien exprimer ce qu'ils pensent ou ce qu'ils sentent' ('the great secret of their art is to express well what they think and what they feel').[31] It was not difficult to imitate the way in which such authors wrote: one had simply to emulate their word order, literary devices or tone and 'l'imitation ne paroîtra point servile' ('the [resulting] imitation will not seem servile'). What is more, the imitator 'ne sera exposé ni à contracter des défauts, ni à perdre de son originalité' ('will be in danger neither of contracting flaws nor of losing his originality').[32] Condorcet implied that the ideas

29. Condorcet, *Eloge historique*, p.35.
30. Condorcet, *Eloge historique*, p.35-36.
31. Condorcet, *Eloge historique*, p.37.
32. Condorcet, *Eloge historique*, p.37.

of these writers were not indistinguishable from their language. Their manner of writing could be copied, but nevertheless used to portray ideas very different from those of the original writer.

Condorcet distinguished between two conceptions of style. The first, as in the example of Voltaire or Fénelon, understood style as an ornament or manner of writing, something that could be borrowed across different discourses. The second conception was the case of authors whose 'style paroît se confondre davantage avec les pensées' ('style appears to be more confused with thoughts'), as if the entirety of the author's mind came with the written language. '[T]els [auteurs] furent Corneille, Bossuet, Montesquieu, Rousseau; tel fut M. de Buffon' ('such [authors] were Corneille, Bossuet, Montesquieu, Rousseau; such was M. de Buffon').[33] The language of these authors could not be detached from their ideas. It was necessary to, and inseparable from, the expression of the underlying essence of the text, which Condorcet described as 'le caractère particulier de l'ame & de l'esprit de l'auteur' ('the specific character of the soul and the mind of the author').[34] In other words – in Buffon's words – 'le style est l'homme même.'

Condorcet seemed to have had the 'Epoques de la nature' in mind as he wrote the *Eloge*, for he described the energy of Buffon's language, experienced by the reader as a form of heat. An imitation of this force would produce only a hollow form lacking true vitality: '[S]i le modèle vous offre des idées originales & grandes, l'imitateur vous présentera des idées communes, chargées d'expressions ordinaires [...]. Le modèle a parlé de tout avec chaleur, parce que son ame étoit toujours agitée; le froid imitateur cachera son indifférence sous des formes passionnées.'[35] If one tries to imitate the words of authors with style, thereby separating language from its inventor and form from content, 'les idées elles-mêmes semblent disparoître' ('the ideas themselves seem to disappear'), and the nature of the work 's'évanouit' ('vanishes'). Authors with this type of profound style 'peuvent être des modèles dangereux. Pour imiter leur style, il faudroit avoir leurs pensées, voir les objets comme ils les voient, sentir comme ils sentent'

33. Condorcet, *Eloge historique*, p.37-38.
34. Condorcet *Eloge historique*, p.38.
35. '[If] the model offers you original and grand ideas, the imitator will only give you common ideas, loaded with ordinary expressions [...]. The model spoke of everything with heat, because his soul was always animated; the cold imitator will hide his indifference under passionate forms.' Condorcet, *Eloge historique*, p.38-39.

('can be dangerous models. In order to imitate their style, it would be necessary to have their thoughts, to see objects as they see them, to feel as they feel').[36] To imitate their style was to 'contracter des défauts' ('contract flaws') as one would a disease, and thus to 'perdre de son originalité' ('lose one's originality').[37] One could become infected both by these authors' words and their invented worlds. To read their language was to be simultaneously and inevitably swept away by the underlying force of their ideas and perceptions. The style of these authors quite literally was the world, or rendering of reality, that it had made, and this harmonization of ideas and language with the structure of the world itself was for Buffon the epitome of natural knowledge. For Condorcet, Buffon's writing was 'dangerous' because, when one truly understood the effect and function of his work, one had no choice but to make worlds as he made them, of his ingredients and energy, and because of his creative, authorial power.

In the eulogy Condorcet identified himself as one of the few who could come away unscathed from reading Buffon. He parodied the natural historian and yet considered himself detached and impartial enough to be able to read Buffon without becoming Buffon. He argued this perhaps because he had understood how to read the *Histoire naturelle* not as a work of science but as a work of literature. Science, from Condorcet's point of view, was the study of facts that became true when ordered and explained through the analytic thought of mathematics. And yet, even in Condorcet's thought, there was an aspect of mathematics that tilted toward the probabilistic and the fictional, that is, toward what could be true. It was evident elsewhere in his corpus, particularly in the 'Dixième époque: des progrès futurs de l'esprit humain' ('Tenth epoch: the future progress of the human mind') of his *Esquisse d'un tableau historique*, which argued, based upon Condorcet's narration of the story of human intellectual history, that the possible could become probable, and even realizable, through the training of the human mind to think, speak and act within the perfect, transparent mathematical language.[38] Similarly the *Histoire*

36. Condorcet, *Eloge historique*, p.38.
37. Condorcet, *Eloge historique*, p.37.
38. For more on Condorcet's ideas of the application of mathematical thought in the social and political realms, see Frank E. Manuel, *The Prophets of Paris* (Cambridge, MA, 1962); Baker, *Condorcet*; and Catherine Kintzler, *Condorcet: l'Instruction publique et la naissance du citoyen* (Paris, 1984). Also see my study of the concept of the invention of an idealized world through mathematical language in the *Esquisse d'un tableau historique des progrès de l'esprit humain* in Hanna Roman,

naturelle, in Condorcet's words a great 'tableau' or ordered system, used the movement between the language of the mind and the language of the world to make readers aware of the development of their thought, training them to act more consciously and deliberately in the future. Buffon's work, he wrote, '[offrait] pour la première fois aux regards des hommes avides de se connoître, & surpris de tout ce qu'ils apprenoient sur eux-mêmes, [...] [la capacité] de retrouver ce qu'ils avoient éprouvé, ce qu'ils avoient vu sans en avoir eu la conscience ou conservé la mémoire'.[39]

Therefore, removing the label of 'scientific' from the *Histoire naturelle* did not mean that the work of an author with a dangerously powerful and compelling style had lost its power and utility. Condorcet wanted to teach future generations how to read Buffon by becoming aware of the mechanics of his language, by understanding his style as an error – but a positive error, a moment in the history of human thought which could be read not necessarily as a failure or a regression but rather productively, as a heuristic tool. Considered this way, Buffon's work represented not the absolute truths of nature but a form of discursive and logical conjecture about how present human activity could reshape the future of the world. It taught readers what it meant to conceive of the world differently and how to render real and productive the latent energy of their thought. In this way, each text produced in the wake of Buffon's style could do something new through the practice of style:

> [V]eut-on [...] prendre pour modèles [des auteurs tels que Buffon], il ne faut point chercher à saisir leur manière, il ne faut point vouloir leur ressembler, mais se pénétrer de leurs beautés, aspirer à produire des beautés égales, s'appliquer comme eux à donner un caractère original à ses productions, sans copier celui qui frappe ou qui séduit dans les leurs.

> [If] one wants [...] to use [authors such as Buffon] as models, one must not strive to capture their manner, one must not want to resemble them, but to become permeated with their beauties, to aspire to produce equal beauties, to try as they do to give an original

'Conjecturing a new world in Condorcet's *Esquisse d'un tableau historique des progrès de l'esprit humain*', *Modern language notes* 129:4 (September 2014), p.780-95.

39. '[O]ffered for the first time to the eyes of people keen to know themselves, and surprised about everything that they learned about themselves, [...] [the ability] to rediscover what they had experienced, what they had seen without having been conscious or having conserved the memory of it'. Condorcet, *Eloge historique*, p.20.

character to one's work without copying that which is striking or seductive in theirs.[40]

The *Histoire naturelle* might in fact become more useful than the writings of the first set of mannerist authors Condorcet listed, because it trained readers not for imitation but for active thinking through the probable and productive realms of literature, realms Enlightenment philosophers believed held the potential for rethinking and changing reality for the better.

Condorcet's *Eloge de M. de Buffon* operated as a work of irony, on two levels: the first of the language *à fleur de peau* and the second of the deeper impact of what was not being directly enunciated. On the surface level Condorcet was sarcastic, making fun of Buffon's scientific errors and exaggerated rhetorical strategies. However, it was through this first, outer level, that the second, hidden level could be accessed. At this level, irony was revealed to be not just a means of mockery but a stylistic element, allowing Condorcet to bring the genre of the eulogy back to its roots as a mastery of rhetoric and hence a mode of literary perception. On the one hand, Condorcet used irony to distance himself from the power of Buffon's style and to critique Buffon as a figure of science. But in the process of transforming Buffon into a literary figure, and the *Histoire naturelle* into a work of *belles-lettres*, he revealed the continuing importance and power of literary style to modify peoples' perception of, and ultimately their action in, the world around them.

Despite his contempt for Buffon, Condorcet devised a way to protect the *Histoire naturelle* from a future of obscurity. He sensed that his generation was increasingly incapable of reading Buffon as he would have been read earlier in the century. Buffon's work was less useful to Condorcet and his contemporaries as a scientific text, but as a literary text it would endure. It was as if the *Eloge de M. de Buffon* were destined for the wrong academy. It was better suited, perhaps purposely so, to Buffon's departure from the Académie française and not the Académie des sciences – for it was the former academy, in Condorcet's mind, to which Buffon really belonged. (Having, like Buffon, belonged to both academies, Condorcet would have fully understood the significance of this gesture.) Perhaps this was Condorcet's last sarcastic insult: that Buffon would not remain a scientist, but that he would remain an author. Yet this claim must also be taken seriously. Buffon's form of

40. Condorcet, *Eloge historique*, p.39.

natural history was no longer seen to be concerned with the real world of facts. It was no longer there to be proven, for whether it was true or false had become beside the point. The worlds invented by the natural historian were probable and verisimilar. They reoriented their readers' perception, teaching how to interact differently and more innovatively with their circumstances.

Condorcet's depiction of Buffon as a literary figure indeed exemplified the development of a scientific language and worldview moving away from Buffon's. However, because Condorcet's voice is not exactly that of modern science, but rather captures a moment between two different ways of thinking, his text serves both as an indication of the future and as a way to comprehend the past. In the *Eloge de M. de Buffon* Condorcet had to use the old in order to invent a new paradigm. He could only articulate a different way of perceiving the natural world through what was already an established standard of scientific thought, that is, through the vocabulary and style of the intellectuals of Buffon's generation. By portraying Buffon's work as a literary object – in the serious discursive sense of the term, as the process of inventing new worlds or visions of reality – Condorcet's *Eloge* came to be an important key for understanding what it meant to study nature in Buffon's day, where language was a part of science itself and not merely a way to express it, and where the literary was thus tantamount to knowledge.

The literary experiment

In nineteenth-century France, Buffon's work provided the discursive tools necessary to continue to translate between the worlds of nature and human thought, especially in the literary movements referred to as 'realism' and 'naturalism'. The tools provided by Buffon's work influenced authors such as Flaubert and Zola, and allowed them to study the moral and social realms of human behavior and interaction according to both the perceived laws of these realms and the energetic, creative power of the authorial voice expressed through language. The issue of the role of the literary voice in the relationship between the world and the text remained crucial. For instance, in Emile Zola's conception of literary thought experiments in *Le Roman expérimental* (1880), the idea of Buffon's hypothesis continued to resonate:

> L'idée d'expérience entraîne avec elle l'idée de modification. Nous partons bien des faits vrais, qui sont notre base indestructible; mais,

> pour montrer le mécanisme des faits, il faut que nous produisions et que nous dirigions les phénomènes; c'est là notre part d'invention, de génie dans l'œuvre. Ainsi [...] je constate dès maintenant que nous devons modifier la nature, sans sortir de la nature, lorsque nous employons dans nos romans la méthode expérimentale [...]. L'écrivain, loin d'être diminué, grandit ici singulièrement [...]. Ainsi donc, au lieu d'enfermer le romancier dans des liens étroits, la méthode expérimentale le laisse à toute son intelligence de penseur et à tout son génie de créateur. Il lui faudra voir, comprendre, inventer.

> The idea of experiment carries with it the idea of modification. We indeed start from true facts, which are our indestructible base; however, to show the mechanism of facts, we must produce and direct phenomena; this is our inventive role, the genius in the work. Thus [...] I note that henceforward, when we use the experimental method in our novels, we must modify nature, without going outside of nature [...]. The writer, far from being diminished, is crucially emphasized [...]. Instead of imprisoning the novelist within narrow boundaries, the experimental method grants him all his intelligence as a thinker and all his genius as a creator. He will have to see, understand, invent.[41]

What Buffon called a hypothesis, Zola now named an experiment, yet similar processes of using language as the basis for understanding reality were at stake in the work of both authors. As Buffon began his project in the initial confusion of observing nature or in the rupture caused by the comet, Zola wrote that the novelist had to 'parti[r] du doute pour arriver à la connaissance absolue' ('start from doubt in order to arrive at absolute knowledge'), continuing to question his observations of society in order to finally discover and explain the fundamental, hidden 'mécanisme de la passion' ('mechanism of passion'). The embodiment and representation of this underlying force in language allowed Zola to describe the novel not only as a clear portrayal of the nature of reality, operating 'selon les lois fixées par la nature' ('according to the laws fixed by nature'), but as an inventive discourse able to 'modify nature' and perhaps even to alter its deterministic course.[42] Zola began the *Roman expérimental* with the idea that the novel should use the 'experimental method' of the sciences, notably medicine, to capture the laws and deterministic

41. Emile Zola, *Le Roman expérimental*, ed. François-Marie Mourad (Paris, 2006), p.55-56.
42. Zola, *Le Roman expérimental*, p.56.

course of social worlds just as this method showed the laws of how organisms adapted and maintained their physiological systems with respect to the natural world. He ended his argument with an affirmation that literature did not imitate the work of the sciences, but rather that, through method, the two realms were able to shape the world not as it was happening but as it would happen: 'la méthode expérimentale, aussi bien dans les lettres que dans les sciences, est en train de déterminer les phénomènes naturels, individuels et sociaux' ('the experimental method, in literature as much as in science, is in the process of determining natural, individual and societal phenomena').[43]

Zola's was but one example of a nineteenth-century literary use of method and hypothesis as both literary and scientific tools – in Zola's case, scientific *because* literary. His understanding of the reordering of intellectual space to make a simulacrum of nature differently useful from nature itself revealed the continuous and underlying operations and assumptions of the relationship of language to knowledge. *Le Roman expérimental* – and indeed Zola's project for the natural and social history of the Rougon-Macquart family – underlines the importance of studying works such as the *Histoire naturelle* as discursive endeavors that make novel use of disciplinary boundaries, not only to better understand Buffon's ideas and their impact but moreover to show how the relationships and interactions between literature and science should always be considered in a nuanced manner.

Returning once more to Buffon: in a well-known article from late in his career about iron ('Du fer', 1783), in the *Histoire naturelle des minéraux*, he reiterated the importance of a strong authorial voice to the discovery and invention of natural knowledge:

> Or les grands procédés de la Nature sont les mêmes en tout, et lorsqu'ils nous paroissent opposés, contraires ou seulement différens, c'est faute de les avoir saisis et vus assez généralement pour les bien comparer [...]. [L]e but du Philosophe naturaliste doit donc être de s'élever assez haut pour pouvoir déduire d'un seul effet général, pris comme cause, tous les effets particuliers; mais pour voir la Nature sous ce grand aspect, il faut l'avoir examinée, étudiée et comparée dans toutes les parties de son immense étendue; assez de génie, beaucoup d'étude, un peu de liberté de penser, sont trois attributs sans lesquels on ne pourra que défigurer la Nature, au lieu de la représenter.

43. Zola, *Le Roman experimental*, p.89.

> Now, the great processes of Nature are the same in all, and when they seem contradictory, contrary or simply different to us, it is because we have not seen and grasped them in a general enough manner to truly be able to compare them [...]. [T]he goal of the natural Philosopher must be to elevate himself high enough to be able to deduce, from a single general effect, understood as a cause, all the particular effects; but in order to see Nature from this great perspective, one had to examine, study and compare it in all the parts of its immense scope; enough genius, a great deal of study, some liberty of thought: these are the three attributes without which one could only disfigure Nature instead of representing it.[44]

Through the ability of the author of natural history to see the world from above, on the level of nature's laws, and to remake this world into words through language, Buffon was able to structure human perception and knowledge according to the same relationships as the objects and events of the real world. Over the course of the article 'Du fer', he showed how the metallic substance of iron was a product of two laboratories: the internal heat of the Earth and the heat of human activity: 'c'est autant la main de l'homme que le feu, qui donne au fer la forme de métal' ('it is as much the hand of man as fire that gives to iron the form of metal').[45] People learned to think and to act in the same way as nature, and the iron produced by their art became indistinguishable from that made by the physical forces of the Earth – and sometimes even more functional than the natural product. The activities of humans and nature were depicted not simply as similar but as necessarily analogous, because of the poetic, energetic power of Buffon's language which generated proportion, harmony and often innovative ways of perceiving between the seemingly distinct realms of thought and reality. The literary, sometimes conjectural, discourse of the *Histoire naturelle* generated a hybrid space between the physical world and the world of ideas. It then fused these two worlds, first through the inventive, energetic voice of its author, and then through the thought experiment of human activity recreating and repurposing this energy as heat in the natural world.

44. Buffon, 'Du fer', in *Histoire naturelle des minéraux*, vol.2 (1783), p.341-496 (344-45).
45. Buffon, 'Du fer', p.354.

Bibliography

Anderson, Wilda C., *Between the library and the laboratory: the language of chemistry in eighteenth-century France* (Baltimore, MD, 1984).

–, *Diderot's dream* (Baltimore, MD, 1990).

–, 'Error in Buffon', *Modern language notes* 114:4 (September 1999), p.691-701.

–, 'Scientific nomenclature and Revolutionary rhetoric', *Rhetorica: a journal of the history of rhetoric* 7.1 (winter 1989), p.45-53.

Aréchiga, Violeta, 'Le feu et la vie: la pensée chimique de Buffon', in *L'Héritage de Buffon*, ed. Marie-Odile Bernez (Dijon, 2009), p.235-44.

Arnauld, Antoine, and Pierre Nicole, *La Logique, ou l'Art de penser*, ed. Charles Jourdain (Paris, 1992).

Ashworth, William B., 'Natural history and the emblematic world-view', in *Reappraisals of the Scientific Revolution*, ed. David C. Lindberg and Robert S. Westman (Cambridge, 1990), p.303-32.

Bachelard, Gaston, *La Formation de l'esprit scientifique: contribution à une psychanalyse de la connaissance objective* (Paris, 1960).

Badinter, Elisabeth, and Robert Badinter, *Condorcet (1743-1794): un intellectuel en politique* (Paris, 1988).

Baere, Benoît de, 'La fiction et l'histoire naturelle au siècle des Lumières: fonctions, enjeux, dangers', *SVEC* 2009:02, p.207-25.

–, 'Une histoire imaginée ... mais vraie: le problème du statut des *Epoques de la nature* de Buffon', in *Les Songes de Clio: fiction et histoire sous l'Ancien Régime*, ed. Sabrina Vervacke, Eric Van der Schueren and Thierry Belleguic (Québec City, 2006), p.247-62.

–, *La Pensée cosmogonique de Buffon: percer la nuit des temps* (Paris, 2004).

Baker, Keith, *Condorcet: from natural philosophy to social mathematics* (Chicago, IL, 1975).

Balavoine, Claudie, 'La *Poétique* de J.-C. Scaliger: pour une mimésis de l'imaginaire', in *La Statue et l'empreinte: la poétique de Scaliger*, ed. C. Balavoine and P. Laurens (Paris, 1986), p.107-29.

Balzac, Honoré de, 'Author's introduction', translated by George Saintsbury, in *The Works of Honoré de Balzac*, ed. William P. Trent (New York, 1900), p.li-lxvii.

–, 'Avant-propos', in *La Comédie humaine*, vol.1 (Paris, 1842), p.7-32.

Beer, Gillian, 'Has nature a future?', in *The Third culture: literature and science*, ed. Elinor S. Shaffer (Berlin, 1998), p.15-27.

Bellos, David, *Is that a fish in your ear? Translation and the meaning of everything* (New York, 2011).

Brown, Theodore M., 'From mechanism to vitalism in eighteenth-century English physiology', *Journal of the history of biology* 7:2 (autumn 1974), p.179-216.

Brunet, Pierre, *L'Introduction des théories de Newton en France au XVIIIe siècle* (Paris, 1931).

Buffon, Georges-Louis Leclerc de, *Un Autre Buffon*, ed. Jacques Roger (Paris, 1977).

–, 'De la lumière, de la chaleur et du feu', in *Histoire naturelle, générale et particulière: supplément*, vol.1 (Paris, Imprimerie royale, 1774), p.1-78.

–, *De l'art d'écrire* (Castelnau-le-Lez, 1992).

–, 'Des époques de la nature', Paris, Muséum national d'histoire naturelle, MS 883.

–, 'Des époques de la nature', in *Histoire naturelle, générale et particulière: supplément*, vol.5 (Paris, Imprimerie royale, 1778), p.1-254.

–, 'Discours sur le style', in *Histoire naturelle, générale et particulière: supplément*, vol.4 (Paris, Imprimerie royale, 1777), p.1-13.

–, 'Du fer', in *Histoire naturelle des minéraux*, vol.2 (Paris, Imprimerie royale, 1783), p.341-496.

–, *The Epochs of nature*, ed. and translated by Anne-Sophie Milon, Jan Zalasiewicz and Mateusz Zalasiewicz (Chicago, IL, 2018).

–, *Les Epoques de la nature*, ed. Jacques Roger, Mémoires du Muséum national d'histoire naturelle, series C, vol.10 (Paris, 1962).

–, 'Essais d'arithmétique morale', in *Histoire naturelle, générale et particulière: supplément*, vol.4 (Paris, Imprimerie royale, 1777), p.46-109.

–, *Histoire naturelle des minéraux*, 5 vols (Paris, Imprimerie royale, 1783-1788).

–, *Histoire naturelle, générale et particulière, avec la description du Cabinet du roy*, 15 vols (Paris, Imprimerie royale, 1749-1767).

–, *Histoire naturelle, générale et particulière, avec la description du Cabinet du roy*, ed. Stéphane Schmitt and Cédric Crémière (Paris, 2007-).

–, *Histoire naturelle, générale et particulière: supplément*, 7 vols (Paris, Imprimerie royale, 1774-1789).

–, 'Notes ayant servi pour la rédaction des œuvres de Buffon', Paris, Bibliothèque du Muséum national d'histoire naturelle, MS 218.

–, 'Réflexions sur la loi de l'attraction', in *Histoire et mémoires de l'Académie royale des sciences*

(Paris, Imprimerie royale, 1749), p.493-500.

–, 'Seconde addition au mémoire qui a pour titre: "Réflexions sur la loi de l'attraction"', in *Histoire et mémoires de l'Académie royale des sciences* (Paris, Imprimerie royale, 1749), p.580-86.

Buffon, Henri Nadault de, *Correspondance inédite de Buffon*, vol.1 (Paris, 1860).

Burnet, Thomas, *Telluris theoria sacra: the theory of the earth, containing an account of the original of the earth, and of all the general changes which it hath already undergone, or is to undergo till the consummation of all things* (London, 1691).

Casini, Paolo, 'Buffon et Newton', in *Buffon 88: actes du colloque international*, ed. Jean-Claude Beaune *et al.* (Paris, 1992), p.299-308.

Cassirer, Ernst, *The Philosophy of the Enlightenment*, translated by Fritz C. A. Koelln and James P. Pettegrove (Princeton, NJ, 1951).

Cavaillé, Jean-Pierre, *Descartes: la fable du monde* (Paris, 1991).

Chassay, Jean-François, *Si la science m'était contée: des savants en littérature* (Paris, 2009).

Cicero, *On the ideal orator (De oratore)*, ed. and translated by James M. May and Jakob Wisse (Oxford, 2001).

Cohn, Norman, *Noah's flood: the Genesis story in western thought* (New Haven, CT, 1996).

Compagnon, Antoine, *Le Démon de la théorie: littérature et sens commun* (Paris, 1998).

Condillac, Etienne Bonnot de, *La Langue des calculs*, ed. Sylvain Auroux and Anne-Marie Chouillet (Lille, 1981).

Condorcet, Marie-Jean Antoine Nicolas, marquis de, *Eloge historique de M. le comte de Buffon* (Paris, Aux Deux Ponts, 1792).

Corsi, Pietro, *The Age of Lamarck: evolutionary theories in France, 1790-1830*, translated by Jonathan Mandelbaum (Berkeley, CA, 1988).

–, 'Buffon sous la Révolution et l'Empire', in *Buffon 88: actes du colloque international*, ed. Jean-Claude Beaune *et al.* (Paris, 1992), p.639-48.

Curtius, Ernst Robert, *European literature and the Latin Middle Ages*, translated by Willard R. Trask (Princeton, NJ, 1973).

Cuvier, Georges, 'Buffon', in *Biographie universelle ancienne et moderne*, 85 vols (Paris, 1812), vol.6, p.234-42.

Dagen, Jean, *L'Histoire de l'esprit humain dans la pensée française de Fontenelle à Condorcet* (Paris, 1977).

Daubenton, Louis-Jean-Marie, *et al.*, 'Première adresse des officiers du Jardin des plantes et du Cabinet d'histoire naturelle', reprinted in Ernst-Théodore Hamy, 'Les derniers jours du Jardin du roi et la fondation du Muséum d'histoire naturelle', in *Centenaire de la fondation du Muséum d'histoire naturelle, 10 juin 1793-10 juin 1893: volume commémoratif publié par les professeurs du Muséum* (Paris, 1893), p.97-100.

Dear, Peter, *Revolutionizing the sciences: European knowledge and its ambitions, 1500-1700* (Princeton, NJ, 2001).

Delille, Jacques, 'Chant premier: la lumière et le feu', in *Les Trois règnes de la nature*, vol.1 (Paris, 1808), p.41-76.

Delon, Michel, *L'Idée d'énergie en France au tournant des Lumières (1770-1820)* (Paris, 1988).

Descartes, René, *Le Monde de M. Descartes, ou le Traité de la lumière et des autres principaux objets des sens* (Paris, Chez Michel Bobin et Nicolas Le Gras, 1664).

–, *Règles pour la direction de l'esprit*, ed. and translated by Jacques Brunschwig (Paris, 1997).

–, *Rules for the direction of the mind*, translated by Laurence J. Lafleur (Indianapolis, IN, 1961).

Dictionnaire de l'Académie française, 4th edn (Paris, 1762), ARTFL *Dictionnaires d'autrefois*, ed. Robert Morrissey, http://artfl-project.uchicago.edu/content/dictionnaires-dautrefois (last accessed March 15, 2018).

Diderot, Denis, 'Pensées sur l'interprétation de la nature', in *Œuvres complètes*, ed. Jean Varloot, vol.9 (Paris, 1981), p.25-111.

Diderot, Denis, and Jean D'Alembert (ed.), *Encyclopédie, ou Dictionnaire raisonné des sciences, des arts et des métiers*, ARTFL *Encyclopédie* project, ed. Robert Morrissey, http://encyclopedie.uchicago.edu/ (last accessed March 14, 2018).

Dieudonné, Jean, *Pour l'honneur de l'esprit humain: les mathématiques aujourd'hui* (Paris, 1987).

Donovan, Arthur, 'Buffon, Lavoisier, and the transformation of French chemistry', in *Buffon 88: actes du colloque international*, ed. Jean-Claude Beaune *et al.* (Paris, 1992), p.387-95.

Dugatkin, Lee Alan, *Mr. Jefferson and the giant moose: natural history in early America* (Chicago, IL, 2009).

Eddy, John, 'Buffon's *Histoire naturelle*: history? A critique of recent interpretations', *Isis* 85:4 (December 1994), p.644-61.

Edelstein, Dan, *The Enlightenment: A Genealogy* (Chicago, IL, 2010).

Ehrard, Jean, *L'Idée de nature en France dans la première moitié du XVIIIe siècle*, 2 vols (Paris, 1963).

Flaubert, Gustave, *Bouvard and Pécuchet*, translated by A. J. Krailsheimer (New York, 1976).

–, *Bouvard et Pécuchet* (Paris, 1945).

Fleck, Ludwik, *Genesis and development of a scientific fact*, translated by Fred Bradley and Thaddeus J. Trenn, ed. Thaddeus J. Trenn and Robert K. Merton (Chicago, IL, 1979).

Foucault, Michel, *The Order of things: an archaeology of the human sciences* (New York, 1970).

France, Peter, 'Lumières, politesse, énergie', in *L'Histoire de la rhétorique dans l'Europe moderne, 1450-1950*, ed. Marc Fumaroli (Paris, 1999), p.945-99.

Furetière, Antoine, *Dictionaire universel* (The Hague, A. et R. Leers, 1690), http://gallica.bnf.fr/ark:/12148/bpt6k50614b (last accessed March 22, 2018).

Galand-Hallyn, Perrine, *Le Reflet des fleurs: description et métalangage poétique d'Homère à la Renaissance* (Geneva, 1994).

Galileo, 'Letter to the Grand Duchess Christina', in *Discoveries and opinions of Galileo*, ed. and translated by Stillman Drake (New York, 1957), p.173-216.

Gayon, Jean, 'De la catégorie de style en histoire des sciences', *Alliage* 26 (1996), n.p.

Goodden, Angelica, *Actio and persuasion: dramatic performance in eighteenth-century France* (Oxford, 1986).

Goodman, Nelson, *Ways of worldmaking* (Indianapolis, IN, 1978).

Gould, Stephen Jay, 'Inventing natural history in style', in *The Lying stones of Marrakech: penultimate reflections in natural history* (New York, 2000), p.75-90.

Goulemot, Jean Marie, *Le Règne de l'histoire: discours historique et révolutions, XVIIe-XVIIIe siècle* (Paris, 1996).

Guerlac, Henry, *Newton on the Continent* (Ithaca, NY, 1981).

Hacking, Ian, *The Emergence of probability* (Cambridge, 1975).

Hahn, Roger, *The Anatomy of a scientific institution: the Paris Academy of Sciences, 1666-1803* (Berkeley, CA, 1971).

Hales, Stephen, *La Statique des végétaux et l'analyse de l'air* [...] *traduit de l'anglais par M. de Buffon, de l'Académie royale des sciences* (Paris, Debure l'aîné, 1735).

–, *Vegetable staticks* (London, W. and J. Innys, 1727).

Hall, Thomas S., *Ideas of life and matter: studies in the history of general physiology, 600 BC-1900 AD*, vol.2 (Chicago, IL, 1969).

–, 'On biological analogs of Newtonian paradigms', *Philosophy of science* 35 (1968), p.6-27.

Hankins, Thomas, *Science and the Enlightenment* (Cambridge, 1985).

Hanks, Lesley, *Buffon avant l''Histoire naturelle'* (Paris, 1966).

Hochstrasser, T. J., *Natural law theories in the early Enlightenment* (Cambridge, 2000).

Hoquet, Thierry, *Buffon: histoire naturelle et philosophie* (Paris, 2005).

–, *Buffon illustré: les gravures de l''Histoire naturelle'* (Paris, 2007).

–, 'History without time: Buffon's *Natural history* as a nonmathematical physique', *Isis* 101 (2010), p.30-61.

Jardine, N., J. A. Secord and E. C. Spary (ed.), *Cultures of natural history* (Cambridge, 1996).

Joly, Bernard, 'La question de la nature du feu dans la chimie de la première moitié du XVIIIe siècle', *Corpus* 36 (1999), p.41-63.

–, 'Les théories du feu de Voltaire et de Mme Du Châtelet', *SVEC* 2001:11, p.212-37.

Jordanova, L. J., 'Introduction', in *Languages of nature: critical essays on science and literature*, ed. L. J. Jordanova (London, 1986), p.15-47.

Kelley, Donald R., 'Between history and system', in *Historia: empiricism and erudition in early modern Europe*, ed. Gianna Pomata and Nancy Siraisi (Cambridge, MA, 2005), p.211-37.

–, ‘Law’, in *The Cambridge history of political thought, 1450-1700*, ed. J. H. Burns and Mark Goldie (Cambridge, 1991), p.66-94.

Kintzler, Catherine, *Condorcet: l’Instruction publique et la naissance du citoyen* (Paris, 1984).

Labio, Catherine, *Origins and the Enlightenment: aesthetic epistemology from Descartes to Kant* (Ithaca, NY, 2004).

Latour, Bruno, *Science in action: how to follow scientists and engineers through society* (Cambridge, 1988).

Leclaire, Lucien, ‘L’*Histoire naturelle des minéraux* ou Buffon géologue universaliste’, in *Buffon 88: actes du colloque international*, ed. Jean-Claude Beaune *et al.* (Paris, 1992), p.355-65.

Lehman, Christine, ‘Gabriel François Venel (1723-1775): sa place dans la chimie française du XVIIIe siècle’, doctoral dissertation, Université de Paris-Nanterre, 2006.

Levacher, Maëlle, *Buffon et ses lecteurs: les complicités de l’‘Histoire naturelle’* (Paris, 2011).

–, ‘Les lieux communs dans l’*Histoire naturelle* de Buffon: rhétorique judiciaire, rivalité dans la récriture et compromis épistémologique’, *Dix-huitième siècle* 42 (2010), p.35-56.

Liebman, Elizabeth, ‘Painting natures: Buffon and the art of the *Histoire naturelle*’, doctoral dissertation, University of Chicago, 2003.

Llana, James, ‘Natural history and the *Encyclopédie*’, *Journal of the history of biology* 33 (2000), p.1-25.

Loveland, Jeff, ‘Another Daubenton, another *Histoire naturelle*’, *Journal of the history of biology* 39 (2006), p.457-91.

–, *Rhetoric and natural history: Buffon in polemical and literary context*, *SVEC* 2001:03.

Macquer, Pierre-Joseph, *Dictionnaire de chimie*, 2 vols (Paris, Lacombe, 1766).

Mairan, Jean-Jacques Dortous de, ‘Nouvelles recherches sur la cause générale du chaud en été et du froid en hiver’, in *Histoire de l’Académie royale des sciences, année MDCCLXV, avec les mémoires de mathématique & de physique pour la même année* (Paris, Imprimerie royale, 1768), p.143-266.

Manuel, Frank, *The Prophets of Paris* (Cambridge, MA, 1962).

–, *The Religion of Isaac Newton: the Freemantle Lectures, 1973* (Oxford, 1974).

Martin, Thierry, ‘Certitude et probabilité selon Buffon’, *L’Enseignement philosophique* 49:3 (January-February 1999), p.5-18.

Mendelsohn, Everett, *Heat and life: the development of the theory of animal heat* (Cambridge, MA, 1964).

Metzger, Hélène, ‘Newton: la théorie de l’émission de la lumière dans la doctrine chimique au XVIIIe siècle’, *Archeion* 11 (1929), p.13-25.

–, *Newton, Stahl, Boerhaave et la doctrine chimique* (Paris, 1974).

–, ‘La théorie du feu d’après Boerhaave’, *Revue philosophique de la France et de l’étranger* 109 (January-June 1930), p.253-85.

Montesquieu, Charles de Secondat, baron de La Brède et de, *De l'esprit des lois*, in *Œuvres complètes*, ed. Roger Caillois, 2 vols (Paris, 1949-1951), vol.2, p.227-995.

–, *Discours sur les motifs qui doivent nous encourager aux sciences*, in *Œuvres complètes*, ed. Roger Caillois, 2 vols (Paris, 1949-1951), vol.1, p.53-57.

–, *Essai sur les causes qui peuvent affecter les esprits et les caractères*, in *Œuvres complètes*, ed. Roger Caillois, 2 vols (Paris, 1949-1951), vol.2, p.39-68.

–, *Lettres persanes*, in *Œuvres complètes*, ed. Roger Caillois, 2 vols (Paris, 1949-1951), vol.1, p.129-373.

–, *Œuvres complètes*, ed. Roger Caillois, 2 vols (Paris, 1949-1951).

–, *The Spirit of the laws*, ed. and translated by Anne M. Cohler, Basia Carolyn Miller and Harold Samuel Stone (Cambridge, 1989).

Mornet, Daniel, *Les Sciences de la nature en France au XVIIIe siècle* (Geneva, 2001).

Morveau, Louis-Bernard Guyton de, Antoine-Laurent Lavoisier, Claude Louis Bertholet and Antoine François, comte de Fourcroy, *Méthode de nomenclature chimique* (Paris, Cuchet, 1787).

Newton, Isaac, *Mathematical principles of natural philosophy and his system of the world*, translated by Andrew Motte, ed. Florian Cajori (Berkeley, CA, 1960).

Ogilvie, Brian, *The Science of describing: natural history in Renaissance Europe* (Chicago, IL, 2008).

Pomata, Gianna, and Nancy Siraisi, 'Introduction', in *Historia: empiricism and erudition in early modern Europe*, ed. Gianna Pomata and Nancy Siraisi (Cambridge, MA, 2005), p.1-40.

Principe, Lawrence M., 'Wilhelm Homberg et la chimie de la lumière', *Methodos: savoirs et textes* 8 (2008), http://methodos.revues.org/1223 (last accessed March 19, 2018), n.p.

Reill, Peter Hanns, 'Buffon and historical thought in Germany and Great Britain', in *Buffon 88: actes du colloque international*, ed. Jean-Claude Beaune *et al.* (Paris, 1992), p.667-79.

–, 'History and the life sciences at the beginning of the nineteenth century: Wilhelm von Humboldt and Leopold von Ranke', in *Leopold von Ranke and the shaping of the historical discipline*, ed. Georg G. Iggers and James M. Powell (Syracuse, NY, 1990), p.21-35.

–, *Vitalizing nature in the Enlightenment* (Berkeley, CA, 2005).

Roche, Daniel, 'Natural history in the academies', in *Cultures of natural history*, ed. N. Jardine, J. A. Secord and E. C. Spary (Cambridge, 1996), p.127-44.

Roger, Jacques, 'Buffon et l'introduction de l'histoire dans l'*Histoire naturelle*', in *Buffon 88: actes du colloque international*, ed. Jean-Claude Beaune *et al.* (Paris, 1992), p.193-205.

–, *Buffon: un philosophe au Jardin du roi* (Paris, 1989).

–, 'Chimie et biologie: des "molécules organiques" de

Buffon à la "physico-chimie" de Lamarck', *History and philosophy of the life sciences* 1 (1979), p.43-64.

–, 'Introduction', in *Les Epoques de la nature*, ed. Jacques Roger, Mémoires du Muséum national d'histoire naturelle, series C, vol.10 (Paris, 1962), p.ix-clii.

–, 'Un manuscrit inédit perdu et retrouvé: les *Anecdotes de la nature* de Nicolas-Antoine Boulanger', *Revue des sciences humaines* 71 (July-September 1953), p.231-54.

–, *Les Sciences de la vie dans la pensée française du XVIIIe siècle* (Paris, 1963).

Roman, Hanna, 'Conjecturing a new world in Condorcet's *Esquisse d'un tableau historique des progrès de l'esprit humain*', *Modern language notes* 129:4 (September 2014), p.780-95.

–, 'Naming as natural process and historical narrative in Buffon's *Histoire naturelle*, 1749-55', *Romance studies* 31:3-4 (November 2013), p.238-50.

Rudwick, Martin, *Bursting the limits of time. the reconstruction of geohistory in the age of revolution* (Chicago, IL, 2005).

Sadrin, Paul, *Nicolas-Antoine Boulanger (1722-1759), ou Avant nous le déluge*, *SVEC* 240 (1986).

Seguin, Maria Susana, *Science et religion dans la pensée française du XVIIIe siècle: le mythe du déluge universel* (Paris, 2001).

Seidengart, Jean, 'Le traitement du problème cosmologique dans l'œuvre de Buffon', in *Buffon 88: actes du colloque international*, ed. Jean-Claude Beaune *et al.* (Paris, 1992), p.309-26.

Sermain, Jean-Paul, *Le Roman jusqu'à la Révolution française* (Paris, 2011).

Shank, J. B., *The Newton wars and the beginning of the French Enlightenment* (Chicago, IL, 2008).

–, '"There was no such thing as the 'Newtonian Revolution', and the French initiated it": eighteenth-century mechanics in France before Maupertuis', *Early science and medicine* 9:3 (2004), p.257-92.

Sloan, Phillip R., 'The Buffon–Linnaeus controversy', *Isis* 67:3 (September 1976), p.356-75.

–, 'Kant on the history of nature: the ambiguous heritage of the critical philosophy for natural history', *Studies in history and philosophy of science, part C: studies in history and philosophy of biological and biomedical sciences* 37:4 (December 2006), p.627-48.

–, 'Natural history, 1670-1802', in *Companion to the history of modern science*, ed. R. C. Olby *et al.* (London, 1990), p.295-313.

Spary, Emma C., 'The "nature" of Enlightenment', in *The Sciences in enlightened Europe*, ed. William Clark, Jan Golinski and Simon Schaffer (Chicago, IL, 1999), p.272-304.

–, *Utopia's garden: French natural history from Old Regime to Revolution* (Chicago, IL, 2000).

Stalnaker, Joanna, 'Painting life, describing death: problems of representation and style in the *Histoire naturelle*', *Studies in eighteenth-century culture* 32 (2003), p.193-227.

–, *The Unfinished Enlightenment: description in the age of the 'Encyclopédie'* (Ithaca, NY, 2010).

Struever, Nancy, 'The conversable world: eighteenth-century transformations of the relation of rhetoric and truth', in *Rhetoric and the pursuit of truth: language change in the seventeenth and eighteenth centuries*, ed. Brian Vickers and Nancy Struever (Los Angeles, CA, 1985), p.79-119.

Svagelski, Jean, *L'Idée de compensation en France, 1750-1850* (Paris, 1981).

Taylor, Kenneth, 'Buffon, Desmarest, and the ordering of geological events in *époques*', in *The Age of the Earth: from 4004 BC to AD 2002*, ed. C. L. Lewis and S. J. Knell (London, 2001), p.39-49.

Terrall, Mary, *The Man who flattened the Earth: Maupertuis and the sciences in the Enlightenment* (Chicago, IL, 2006).

Viala, Alain, *La Naissance de l'écrivain: sociologie de la littérature à l'âge classique* (Paris, 1985).

Vila, Anne C., *Enlightenment and pathology: sensibility in the literature and medicine of eighteenth-century France* (Baltimore, MD, 1998).

Volpilhac-Auger, Catherine (ed.), *Dictionnaire Montesquieu* (Lyon, 2013), http://dictionnaire-montesquieu.ens-lyon.fr/fr/article/1376426324/en (last accessed March 19, 2018).

Weil, Françoise, 'La correspondance Buffon–Cramer', *Revue d'histoire des sciences et de leurs applications* 14:2 (1961), p.97-136.

Wessely, Anna, 'Transposing "style" from the history of art to the history of science', *Science in context* 4 (1991), p.265-78.

Westfall, Richard S., 'Hales, Stephen', in *Catalogue of the scientific community of the 16th and 17th centuries*, http://users.clas.ufl.edu/ufhatch/pages/03-Sci-Rev/SCI-REV-Home/resource-ref-read/major-minor-ind/westfall-dsb/index.htm (last accessed March 19, 2018), n.p.

–, 'The rise of science and the decline of Orthodox Christianity: a study of Kepler, Descartes, and Newton', in *God and nature: historical essays on the encounter between Christianity and science*, ed. David C. Lindberg and Ronald L. Numbers (Berkeley, CA, 1986), p.218-37.

Wolloch, Nathaniel, *History and nature in the Enlightenment: praise of the mastery of nature in eighteenth-century historical literature* (Farnham, 2011).

Yeo, Richard, 'Classifying the sciences', in *The Cambridge history of science*, vol.4: *Eighteenth-century science*, ed. Roy Porter (Cambridge, 2003), p.241-66.

Zola, Emile, *Le Roman expérimental*, ed. François-Marie Mourad (Paris, 2006).

Index